The Story of Man's First Station in Space

by **William J. Cromie**

David McKay Company, Inc.
New York

Book Design by George H. Buehler

Library of Congress Cataloging in Publication Data

Cromie, William J
Skylab.

Includes index.
SUMMARY: Describes the experiences of accomplishments of the astronaut crews that participated in the
three separate and increasingly longer missions living
and working in the Skylab space station.
1. Skylab Project—Juvenile literature. [1. Skylab
Project. 2. Space Stations] I. Title.
TL789.8.U6S555   629.44′5   74–25983
ISBN 0-679-20300-1

MANUFACTURED IN THE UNITED STATES OF AMERICA

Contents

Introduction

When I was ten, I used to read about the adventures of a fabulous space character named Buck Rogers. Buck wore a rocket-powered backpack that made him a one-man space ship. In tight situations, he whipped out his space gun. A squeeze of the trigger propelled him out of danger. He rocketed from planet to planet, the way we go from suburb to suburb, and battled the deadly Catmen of Mars on the way.

That cartoon strip was set in the twenty-fifth century. I never believed I would live to see any of the marvels it depicted. Yet only twenty-one years later, in 1961, I was writing about a thirty-eight-year-old test pilot named Alan Shepard, who squeezed into a cramped, tinny-looking man-can on top of an 83-foot-tall Redstone rocket. From miles away, I watched as the rocket exploded into life and sent that modern Buck Rogers into space. The sight raised goose pimples on my flesh and brought tears to my eyes. Since that time, I have been at Cape Canaveral every time a man left Earth on the tip of a gleaming rocket with a tail of fire. I have never been bored with the sight, and I still get goose pimples.

Five years before Buck Rogers was born in a 1929 comic strip, a twelve-year-old named Wernher von Braun strapped rockets to his coaster wagon and whizzed down a Berlin street. While I was reading about rockets in the twenty-fifth century, von Braun was building them. He was in charge of constructing both the V-2 rocket that shattered London and Antwerp in World War II, and the Redstone booster that made Alan Shepard the first American in space.

In 1966, I sat in von Braun's office at the George C. Marshall Space Flight Center in Huntsville, Alabama, and talked about the huge Saturn 5 rocket his engineers had just built. It was designed to break man free from the gravitational confines of Earth for the first time and send him to the moon. It did that in 1968 when Frank Borman, James Lovell, and William Anders spent Christmas Eve orbiting the moon.

The eleven years from Shepard's flight to the last lunar landing in 1972 included some of man's greatest adventurers. Yet they were only the beginning—short expeditions to explore the unknown ocean of space. In 1973-74, man began living and working in space, as opposed to exploring a part of it then quickly returning to Earth. During the Skylab program, men spent one month, then two months, then almost three months living continuously in space. It was man's first home in space.

During the 172 days that they lived there, the astronauts donned backpacks and flew around the inside of the space station the way Buck Rogers flew around the space globes of the Martian Catmen. They tested guns that sent them careening around their space house. Before the end of this century, it is likely astronauts will use such transport to reach "space globes"—vehicles that we call artificial satellites today.

Skylab proved not only that man could survive for long periods in space but could do most of the things he does on Earth. Time and again, the missions were plagued with problems of every kind—rocket malfunctions, sickness, equipment breakdowns, mistakes. Time and again, astronauts solved the problems, even though the solutions involved incredible difficulties such as a six-hour spacewalk or repairs to the outside of a space station never before attempted. For hours astronauts hung upside down outside the station, doing repair work as Skylab orbited 240 miles above Earth. At other times, they worked at "routine" tasks such as photographing gigantic storms on the Sun, making new discoveries about Earth, or doing experiments that someday may lead to the first space factories and the first large space-built buildings. Nineteen of the experiments performed by the astronauts were designed by high school students. They included things such as looking for a new planet, predicting volcanic eruptions, and determining if spiders can spin webs and fish can swim straight when they can't tell up from down.

The one-man Mercury flights begun by Shepard, the two-man Gemini flights, and the three-man journeys to the Moon opened the door to space. They compare to the voyages of the Vikings and Christopher Columbus. Skylab showed that man could pass through the door and live on the other side. It was like establishing the first colony in the New World. Skylab was the first colony in space.

During Skylab, I lived near the Johnson Manned Spacecraft Center, outside Houston, Texas, and I wrote about the day-to-day triumphs and trials. After working with them since 1961, many astronauts were old friends. I talked with them before they went on each mission and when they returned. During the flights, I listened to their words as they were telemetered from space. Those I did not hear, I read in the thousands of pages of transcripts prepared by the National Aeronautics and Space Administration. I attended dozens of briefings, and press conferences, and read thousands of pages of technical reports. These were the sources of information for this book. As much as possible I have used the actual words of the astronauts and the men that supported them on the ground. Skylab was a great achievement for man, and a great experience for me. I wrote this book to share the drama, glory, and flavor of it with others.

I especially wanted to share it with young people. Many of you were born during the space age. For you, space flight has always been a reality. You did not experience the awe and wonder of witnessing the swift transition from propeller planes, barely able to cross the ocean, to space ships that land on the Moon. For you, Buck Rogers always was real.

But I hope you will experience awe and wonder, too, as you witness space flights of the future. The exploration is not over. Man will land on Mars; probably in your lifetime. You may see him go even further, possibly to the moons of Jupiter. Even more exciting—the people who go into space in your lifetime won't all be astronauts. They will be scientists, engineers, physicians, and technicans. They will be women as well as men. As a reader of Buck Rogers, I could only dream of going into space. As a reader of *Skylab,* you can achieve it.

Accident in Space

Hot orange flames burst from five bell-shaped nozzles at the bottom of the huge rocket. The heat could turn metal to liquid, and the roar deafened people watching three miles away. Fire, smoke, and thunder continued for four seconds, but the rocket did not move. Pumps working with the strength of thirty Diesel locomotives each second forced almost fifteen tons of liquid oxygen and kerosene into the five engines. When the thrust, or upward push, reached 7,700,000 pounds, holddown arms released the rocket and it began to rise slowly.

Taller than a thirty-story building, the Saturn V rocket gradually picked up speed as it rose from launch pad 39A at the Kennedy Space Center on Florida's east coast. Then it began to tip away from the launch tower and arch out over the Atlantic Ocean. Cheers went up from crowds of "bird watchers" who had come to see the lift-off. Many shouted, "Go, go, go!"

The Saturn rocket was the same kind that had sent crews of three astronauts to the Moon. For those journeys, three rocket stages, each with their own fuel tanks and engines, boosted the Apollo spacecraft into Earth orbit, then gave it a powerful, fiery push to the Moon. Now only two stages were full of fuel. The third—48 feet long and 22 feet in diameter—had been outfitted as a spacehouse called "Skylab." Divided into two "stories," it contained as much room as a small three-bedroom house, and it included living quarters, a kitchen and a workshop. Attached to one end were a mount for telescopes, a place to allow spacecraft to dock with the station, and an airlock so astronauts could go back

These three astronauts were named by the National Aeronautics and Space Administration as the prime crew of the first Skylab mission. They are (left to right) Joseph P. Kerwin, science pilot; Charles Conrad, Jr., commander; and Paul J. Weitz, pilot.

and forth into space. Skylab would be parked in an orbit 270 miles high, and three crews of astronauts would take turns living there. The first crew was to be launched the next day—May 15, 1973.

The launch of the empty spacehouse had been perfect. The happy crowd of well-wishers could not see that trouble began even as they strained to follow the flaming rocket across the pale blue sky. As speed built up, so did the pressure of air which flowed into openings under a thin aluminum sheet designed to protect Skylab against tiny meteorites. This shield lay over the top of the spacehouse, like a roof covering. At 63 seconds after launch, air pressure forced its front edge away from the spacecraft. Then the supersonic flow of air past the rocket ripped the aluminum shield away.

The broken shield smashed into "wings" on the sides of Skylab. Folded back against the sides of the space station during launch, they were supposed to swing out like the wings of a bird once the craft reached orbit. Each wing contained thousands of solar cells which convert heat from the Sun into electricity to power Skylab. As it tore away, the shield ripped off tiedowns holding the right wing, and it opened partially. Then exhaust flames from the second stage of the rocket hit the crippled wing and burned it away. A straplike piece of metal from the shield wrapped around the other wing and jammed it against the side of Skylab. The huge bird that had proudly soared away from Cape Canaveral, Florida, just ten minutes before was now a crippled goose without wings and with a big scar on its back.

Skylab Reaches Orbit

On the ground, astronauts and flight controllers were not aware of the problems. The astronaut crew included Navy test pilot Charles "Pete" Conrad of Philadelphia, a veteran of a landing on the Moon; Navy flight surgeon Joseph P. Kerwin, the first medical doctor assigned to a U.S. spaceflight, and Paul J. Weitz of Erie, Pennsylvania, a Navy pilot and veteran of combat flying in the Vietnam War. This would be Conrad's fourth space mission; it would be the first for Kerwin and Weitz. They watched the launch from a secluded spot with other astronauts and were elated with what they saw.

The crew went back to preparing for their own launch next day. As they did, engineers at the Mission Control Center near Houston, Texas, 1,200 miles away, picked up the first signs of trouble. Tiny sensors on the meteoroid shield showed that it vibrated badly. Then the signals stopped, indicating that the

Skylab, the United States' first manned space station, was 118 feet long and contained the same volume as a moderate two-bedroom house. Its two sets of solar energy panels were designed to develop enough electrical power to supply four average homes, but one panel was lost soon after launch.

shield had ripped away. However, the space station went into orbit exactly as planned.

The next step was for the nose of the space station to split apart and be blown away. Fifty-six feet long, this 26,000-pound nose was made of four aluminum shells. These fit over the telescope mount and two compartments, or modules, on the front end of Skylab. The shells formed a protective cover and made a smooth, tapered shape that enabled the whole assembly to fly better. Five minutes after reaching orbit, a small explosive charge blew these clear of the spacecraft.

Exposed at the front sat a telescope mount, 15 feet tall and weighing 24,656 pounds. Upon a signal from the Houston control center, this mount rotated 90 degrees on a rigid trusswork. When locked in position, it straddled the top of the other modules on the front of Skylab.

More signals from mission control, and four wings unfolded, accordion-fashion, from the telescope mount. Locked in position, they resembled the blades of a windmill, 102 feet from tip to tip. These wings held a total of 164,160 solar cells. These cells converted energy from the Sun into 10,500 watts of electricity, about half the "juice" needed to power Skylab. The rest of the power was supposed to come from cells on the two crippled wings along the sides of the workshop and living quarters.

Forty-one minutes after the launch, signals were sent to open this second set of wings. Nothing happened. Thirty minutes later, the station—moving 17,500 miles an hour—was over Australia. The signal was sent again. Again the wings did not respond. As Skylab finished its first trip around the world and passed over Texas, flight controllers signaled a third time. By then everyone was sure that the ambitious plan to keep three crews of astronauts in an elaborate space-house for a total of 140 days was in serious trouble.

Yet, worse was still to come. In addition to serving as a barrier against meteroids which might puncture holes in the spacehouse, the aluminum shield acted as an insulator. Special paint on it limited the amount of heat that would be absorbed. Without this protection, temperatures inside and outside Skylab began to rise.

That night, over a "bon voyage" steak dinner, the astronauts were given the bad news. Their reservations for a stay in the spacehouse had been canceled. William C. Schneider, Skylab program director, announced that their launch would be delayed until the following Sunday, May 20. He also told them that their planned 28-day stay in space—twice as long as any previous manned flight—might be cut to 21 days or less. "We're not trying to bury the patient yet, but we know it's sick and in the hospital," Schneider said gloomily.

This "patient" was the key element in a $2.6 billion plan. It was an ambitious sequel to the spectacular Moon landings. The plan aimed at doubling the time U.S. astronauts spent in space—from 14 to 28 days. Two months after

Conrad, Kerwin, and Weitz returned, another crew would live in Skylab for 56 days. This would again double the time men had lived in space. Finally, in October, a third crew would complete another mission, which was scheduled for 56 days, but some people thought it might be extended to 70 days, or even longer.

Cooling It

The worst problem turned out to be overheating. Officials of the National Aeronautics and Space Administration (NASA) calculated that they could get enough power for some sort of a mission by using fuel cells and batteries on the Apollo spacecraft which would ferry the crew to Skylab. But without protection from the Sun—which shines much more fiercely in space than on Earth—the inside of the spacehouse grew hotter and hotter. As temperatures climbed as high as 190° F., controllers knew it would be impossible for astronauts to live and work aboard Skylab.

They also worried that the heat would spoil food stored aboard, fog sensitive film, and ruin drugs in the medical kits. Heat on the outside metal of Skylab rose to 325° F. Some engineers feared this might cause the metal to buckle and tear. High temperatures also could cause the breakdown of Styrofoam insulation in the space station walls, and this would release gases that might poison the astronauts.

First priority was finding a way to cool Skylab quickly. Engineers in Houston sent radio signals to fire small thrust rockets aboard the space station and turn the exposed part away from the Sun. After two days of work, controllers decided to keep the 85-foot-long station tilted up at a 50-degree angle. This shaded the exposed part. At the same time, solar panels on the telescope mount received enough sunlight to generate power. The plan worked. Temperatures dropped from an average of 130° F. to about 105° F. This was still hot, but it gave the people on the ground time to figure out a way to replace the lost shield.

NASA's top officials ordered all available engineers and technicians to "brainstorm" a solution. One proposal involved a giant balloon to cast a shadow over Skylab. The 42-foot balloon would be carried up uninflated to the station by the astronauts. Once in position on the outside of Skylab, it could be inflated. The trouble with this idea was that no one had a balloon 42 feet long and 11 feet wide. It would have to be specially made, and that would take too much time.

NASA engineers, private contractors working on Skylab, and companies that had nothing to do with the project, proposed various shades, sails and umbrellalike devices. As the May 20 launch date approached, however, it became apparent that engineers could not make a good sun shield in time. Skylab had cooled enough so that the launch could be delayed again without risking damage to the spacehouse or the supplies aboard. Officials set May 25 as the new launch date.

Working overtime and on a volunteer basis, workers at the Johnson Space Center (JSC) near Houston invented a sail-like shield made of an aluminumized fabric. It could be put in place by astronauts leaning out of the hatch of the Apollo spacecraft. Weitz or Kerwin would attach the rolled-up shield to Skylab, while Conrad held the spacecraft in position. The 20-foot-long, 10-foot-wide shield would then be unfurled like a sail.

Meanwhile, engineers at the Marshall Spaceflight Center in Huntsville, Alabama, came up with another good device. This consisted of a 20-by-20-foot foillike shade mounted on two poles with an "A" shape. The astronauts would extend the metal poles along the roof of the living and working quarters. Then the shade could be stretched between the poles.

While working on their sail-shield, engineers at JSC asked an imaginative fellow named Jack A. Kinzler to help them make some parts. Kinzler, astronaut Conrad's next-door neighbor, was chief of the Technical Services Division. After studying the problem, he began designing his own sunshade, one that astronauts could put up without getting out of their spacecraft. It was a 22-by-24-foot aluminumized nylon parasol. It could be folded to fit inside a canister 53 inches long. The astronauts could place this canister inside the inner door of a small airlock. This was a two-doored channel through the spacehouse wall. It was used for moving scientific experiments to the outside of Skylab and back. With the canister in place, the inner door of the airlock would be shut and the outer door opened. Using a pole that passed through a special seal on the inner door, the astronauts would shove the parasol to the outside. Once its four telescoping "ribs" cleared the outside of the space station, springs would snap open the bright orange umbrella.

Kinzler built the first model of his space umbrella using four telescoping fishing poles for ribs. When Pete Conrad and the director of JSC, Christopher Columbus Kraft, saw it demonstrated, they ordered Kinzler to make one that could be used in space. Work on it progressed so fast that flight directors picked the parasol as the best device for the astronauts to try first. It eliminated the need for astronauts to work outside, and, if something went wrong, they could get rid of it and try one of the other shades. Working on a twenty-four-hour schedule, Kinzler and his men did six months work in six days. They delivered their heat shield to Cape Kennedy * only hours before NASA scheduled the astronauts to blast off.

Bold Rescue Is Planned

Sure that one of the three sunshades would work, the top men in NASA conceived a bold plan: The three astronauts would blast off from Cape Kennedy

---

* The name of the point of land on Florida's east coast where U.S. manned spacecraft are launched was changed from Cape Canaveral to Cape Kennedy in 1963. The change honored former President John F. Kennedy. Residents of the area protested the new name, and NASA changed it back to Cape Canaveral in 1974. The name of the launch facility, however, has remained the Kennedy Space Center. Skylab flights were launched from the Kennedy Space Center while it was on Cape Kennedy.

on the morning of May 25. After a seven-hour chase, they would catch up with the space station about 4:00 P.M. The three would dock with Skylab and eat a brief meal. About 6:00 P.M., they would undock and maneuver over to the jammed wing. One astronaut would hang out the open hatch of the Apollo spacecraft while the other held his heels. In this position, Kerwin and Weitz would take turns trying to free the jammed solar panel with long-handled cutting tools, hooks, and pry-bars. When the wing was freed, they would redock, eat a second meal, and try to get a night's sleep. Next day, the three planned to enter the sweltering space station and rig the big orange umbrella over the roof. Should that fail, they would don their space suits and attempt to put up the twin-pole shade. If that didn't work, they would try to unfurl the JSC sail.

Nothing like this had ever been thought of, much less attempted, in the history of the space program. It also was a difficult and dangerous plan. Conrad, Kerwin, and Weitz would have to perform the first major repair job in space. They had received only ten days training. All the procedures and tools were new to them.

A great deal was at stake. If they failed, a $200-million space station would become a useless hunk of junk. An ambitious eight-month-long, $2.6-billion space effort, in planning since 1965, would be scrapped. The failure might cause an angry Congress to cut off money for future manned spaceflights.

If they succeeded, there would be no doubt that man belonged in space. Arguments that anything men could do in space, machines could do cheaper, would lose their effectiveness.

As the launch day approached, the gloom of ten days before turned into high hopes and confidence. Could it really be done? reporters asked Christopher Kraft.

"I believe it can be done," he answered. "If I didn't, I wouldn't let them go."

Repairs in Space

"We fix anything."—Pete Conrad

The astronauts had been chasing the damaged space station since 8 A.M. of May 25. It was now 3:30 in the afternoon and they were flying their Apollo spacecraft 259 miles above the Pacific Ocean.

"Tallyno! The Skylab!" Pete Conrad's happy voice crackled over the radio. The tracking station on Guam heard his words and sent them on to the Mission Control Center near Houston.

Pete saw sunlight glinting off the Skylab like a star so bright it shone in broad daylight. The astronauts overtook the space station at the rate of 20 miles an hour.

As the station grew larger in the windows, Paul Weitz thought of their near perfect lift-off from Cape Kennedy that morning. "Boy, that sure was a smooth ride," he said. Weitz, almost forty-one years old and a space rookie, had expected much more vibration.

Joe Kerwin, also forty-one, sat in the middle seat between Weitz and Conrad. Conrad would celebrate his forty-third birthday during this mission. Kerwin had waited eight years as a scientist-astronaut to get on a spaceflight. Now, as he thought of the blast-off that morning and saw Skylab up ahead, the Navy doctor felt the wait was well worth it.

true for things like turkey and gravy and chicken and gravy. Dinner went pretty well, except Paul found a tree trunk in his asparagus."

A Tough Task

Dinner over, the astronauts tackled the first repair job ever attempted in space. Pete undocked the spacecraft and eased it up to the damaged solar wing. All three put on space helmets and gloves, then checked for leaks in their spacesuits. The air was let out of the spaceship and the hatch opened to the deadly vacuum of space.

Weitz stood up in the hatchway to his waist. With Kerwin holding onto his legs, he leaned out. Using a long-handled tool, Paul attempted to pry away the metal holding the wing against the side of Skylab.

Here is what the astronauts said over the radio as they worked on freeing the winglike panel.

Weitz: That son-of-a—— is poked in there like it's nailed in.

Conrad: Can you pry it up?

Weitz: I don't have that much control of the tool. Have you got hold of my legs, Joe?

Kerwin: One of them.

Conrad *(to the ground):* A tiny metal strap . . . the screws in it have riveted into the solar panel. We pulled as hard as we could, but we can't get it out.

Houston Control Center: You've got about fourteen minutes to sunset. *(The two craft would fly into darkness, and it would not be light enough to continue with the work.)*

Closeup view of damaged and partially deployed OWS solar array wing, showing the aluminum strapping which prevented the solar panel from deploying properly. The huge bird that had proudly soared away from Cape Canaveral was now a crippled goose.

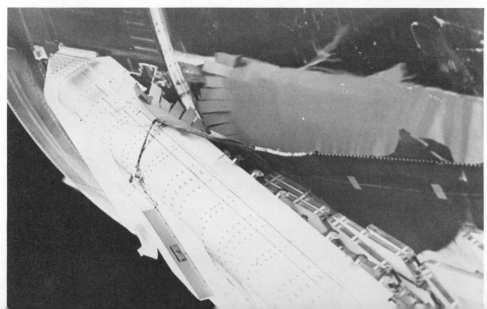

Weitz: I hate to say it, but we ain't going to do it with the tools we got.

Conrad: Okay, Houston, we're going to have to give up. I really feel bad because it's just a tiny half-inch strap. But, boy, did it rivet itself to the side of the panel.

Weitz pulled on the metal so hard he moved the two spacecraft together. Pete saw this and tried to back off. An automatic system aboard Skylab sensed the movement, and started to maneuver it away from the Apollo craft.

Weitz was stretched between the two, like someone with his feet in a boat trying to hold on to a dock as the boat moved away.

"When I saw what was happening, I started to move back in again," Conrad explained later. "We had a little in-and-out and a little hollering and pulling."

"The tool was caught," Weitz added. "I wasn't about to hang on to that thing."

The repair attempt was given up about 7:00 P.M. Conrad maneuvered the Apollo back around to the front of the station, but discovered to his horror that he couldn't get the two craft latched together again.

Trying to Get Together

Pete made two attempts to dock with Skylab, but the spacecraft bounced off both times. The commander tried a third time in the darkness, but this, too, failed. Ground control then suggested using more of the jet thrusters on the spacecraft to force it into position, but this didn't work either.

When further attempts produced no success, the crew put their helmets and gloves back on and let the air out of the spacecraft again. They removed a hatch in the nose of Apollo so they could get at the probe and latch mechanism. Some rewiring was done, and the spacecraft was backed off for one last attempt at docking.

Instead of relying on the probe and latch system to bring the craft together those last few inches, Pete used the spacecraft's small rocket thrusters to shove them together. "I really poured the coals to it," he said later.

As Skylab came within radio range of a tracking ship in the Atlantic Ocean at 10:50 P.M. that night, ground controllers heard Conrad report with relief in his voice: "We've got a hard dock out of it!" The two craft were securely latched together.

"Hey, way to go. Good show," came the reply.

A cheer went up at the Mission Control Center.

Boarding the Space Station

The linked spacecraft now circled Earth every hour and a half in an orbit from 270 to 275 miles high. As the astronauts slept, ground teams sent up signals to begin automatically building up oxygen and nitrogen pressures in the space

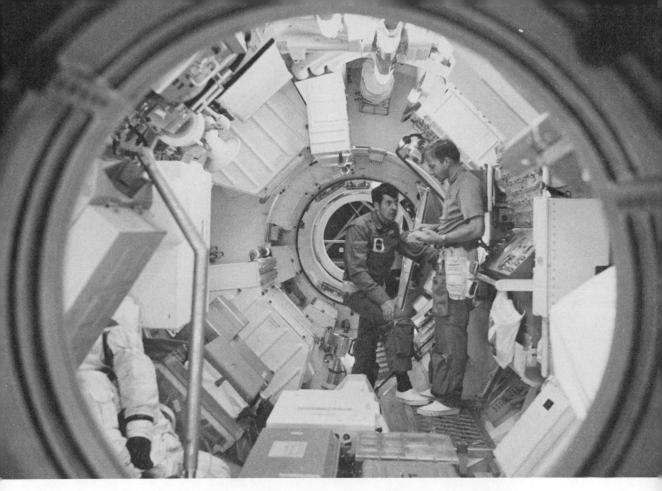

Two members of the prime crew, Joseph P. Kerwin (left) and Paul J. Weitz, are seen here in the Multiple Docking Adapter trainer during an eleven-day simulation schedule prior to the launching of Skylab.

bulky spacesuits. Joe Kerwin stayed in the Apollo spacecraft to look out the window and report on the opening and positioning of the parasol.

Raising the Sunshade

The orange sunshade had been carefully folded by parachute riggers and squeezed into an 8-inch-square, 53-inch-long canister. The astronauts slid this box into one of two small airlock openings through the wall of the workshop. They closed the inner door of the airlock and opened the outer one. The handle, or main shaft, of the parasol extended throught the inner door. Conrad and Weitz slowly pushed the umbrellalike device beyond the outer wall by means of this shaft, screwing on extra rods as they did.

About 6:30, Conrad reported to the ground that "the rod extension has gone easily. It's pretty warm down here, so we are progressing slowly and taking little heat breaks."

The edge of the parasol, folded up like a closed beach umbrella, slowly approached the outer wall of the space station. As soon as the four telescoping umbrella ribs were in the clear, powerful springs snapped them out to an extended position. The parasol was supposed to open up to a 24-foot-long, 22-foot-wide rectangular sunshade.

This happened while Skylab orbited out of radio contact with the Earth. At the Mission Control Center, flight controllers and those who worked on the parasol waited anxiously to hear if the device had worked.

Just after 8:00 P.M., Conrad told the tense and nervous people on Earth: "We had a clear deployment."

Cheers went up.

But the astronaut added: "It's not laid out the way it's supposed to be."

"The two ribs closest to the Apollo spacecraft came up smartly," added Kerwin, "but it doesn't look like the back ones came up all the way."

Three large wrinkles in the after part of the parasol prevented it from opening all the way. Conrad and Weitz moved the parasol in and out rapidly with the center shaft and rotated it, trying to shake the wrinkles out.

The orange nylon was above the wall of the workshop because it had to be pushed out about twenty-one feet from the wall before the ribs could swing up and out. Engineers and controllers figured that the folds could be straightened out by pulling the parasol down against the surface of the workshop.

About 8:30 they radioed the astronauts: "Once we get the parasol pulled back in close to the workshop, we do not plan to do anything more with it tonight."

Stretching the parasol over the surface of Skylab partially smoothed out the wrinkles. Then exposure to the heat of the Sun gradually completed the job.

At 10:30 P.M. William Schneider, Skylab Program Director, told weary reporters covering the flight: "The deployment of the sunshield has been successful. It looks like things are turning in our favor."

Neil Hutchinson, one of the flight directors, explained: "Temperatures on the outside skin of the spacecraft have dropped fifty or sixty degrees in the first hour and a half that the parasol has been out. We hope that by tomorrow, the inside of Skylab will be like Phoenix [Arizona] on a warm, sunny day. It was a busy and fruitful day. We are now looking forward to a very successful twenty-eight-day mission."

Chapter

A Dangerous
Spacewalk

The $2.6 billion Skylab program had been saved by a $75,000 golden orange parasol. As temperatures dropped rapidly in the space station, officials cleared the astronauts to fly for the full 28 days of the mission. Moreover, successful spreading of the sunshade paved the way for two additional missions.

As originally planned, the first crew would double the U.S. space endurance record established in 1965. In December of that year, Frank Borman and James A. Lovell, Jr., orbited Earth for 14 days. Three Russian cosmonauts stayed in space for 24 days in 1971, but they died during their return to Earth. The deaths were caused by air leaking through a faulty valve and not by medical problems. Therefore doctors figured that U.S. astronauts could safely stay up for 28 days.

If Conrad, Kerwin, and Weitz could survive and do difficult tasks for that period, a second crew would attempt to man Skylab for 56 days. This would double man's time in space again. If this second crew experienced no difficulties, a third crew would move into the space house for another 56 days. As it turned out, later, the second and third crews actually stayed in space longer than planned.

Setting Up Housekeeping

The next job facing Conrad's fixit team was to unpack and set up housekeeping. This began Sunday morning, May 27. Lights, air-conditioning, air cleaners, water, toilets, and scientific equipment were turned on and checked.

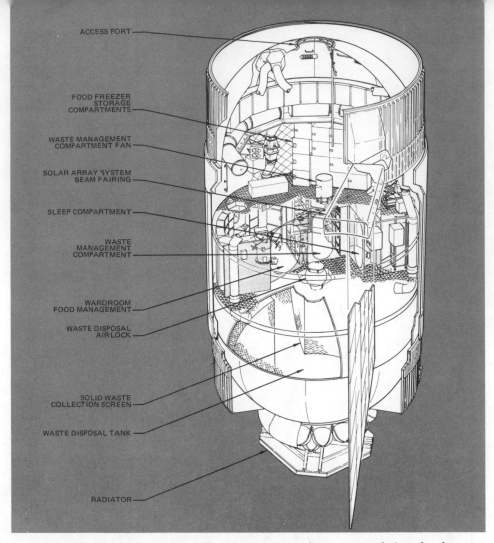

ACCESS PORT

FOOD FREEZER
STORAGE
COMPARTMENTS

WASTE MANAGEMENT
COMPARTMENT FAN

SOLAR ARRAY SYSTEM
BEAM FAIRING

SLEEP COMPARTMENT

WASTE
MANAGEMENT
COMPARTMENT

WARDROOM
FOOD MANAGEMENT

WASTE DISPOSAL
AIRLOCK

SOLID WASTE
COLLECTION SCREEN

WASTE DISPOSAL TANK

RADIATOR

The Orbital Workshop is a two-floor structure providing accommodations for the crew and a primary experiment area.

To prevent equipment from falling into a jumble during launch and boost into orbit, technicians packed everything in boxes and lashed the boxes to the floors of the space station. These had to be unpacked and all items stored away. Keeping three men alive and doing experiments for at least 140 days required some 20,000 items, including everything from 2,000 pounds of food to 1,200 aspirin.

Temperatures still remained close to 100° F., so unpacking was hot work. The astronauts were surprised at how easy it was to handle large, bulky boxes in weightlessness.

"Moving big articles is no problem. Most of the food boxes can be handled by one man, so this part of the job went quicker than expected," Conrad noted.

Conrad gave a vivid description of what life was like those first days in the space station: "It's taking awhile to get the hang of things," he told ground controllers. "The water system [used with powered food] has gas in it, so food handling takes longer than when we practiced on Earth. A bag with corn in it broke when I was inflating it with water. I got corn all over everywhere.''

In weightlessness, all objects float around freely. To hold themselves in place the astronauts wore special shoes. Triangular cleats on these shoes fit into openings on the gridwork floors. An astronaut could lock himself in place by putting the cleats in the openings and turning his foot slightly. Also, a spaceman could slip his foot into toe straps, or restraints, fixed at various locations.

"There are places where not having restraint is bothering us," Conrad reported. He, Kerwin, and Weitz couldn't hold themselves in place when they wanted to use the toilet. "Your feet won't stay under the straps by the toilet when you have triangular shoes on," the Skylab commander explained. "You slip out of them, so we're bouncing off the walls in there."

Drawing Blood

On Sunday night, May 27, the crew slept in the multiple docking adapter where temperatures were a comfortable 68–69°. Joe Kerwin took blood samples from himself and his two space mates next morning. Upon return to Earth, the samples would be examined to determine what effects a long period of weightlessness has on blood and on the body's resistance to disease.

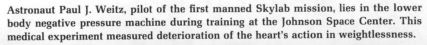

Astronaut Paul J. Weitz, pilot of the first manned Skylab mission, lies in the lower body negative pressure machine during training at the Johnson Space Center. This medical experiment measured deterioration of the heart's action in weightlessness.

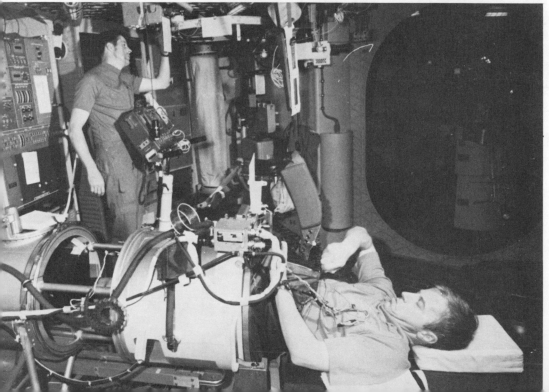

Next day, the three ate breakfast for the first time in the wardroom, or dining area, on the lower level of the workshop. Temperatures there had dropped into the 80s.

In addition to moving in, the crew started medical tests on themselves Monday. Paul Weitz slid up to his waist into a gadget which resembled an oil drum lying on its side. A seal on the opening fit tightly around his waist. Pressure inside the drum was dropped lower and lower, and blood started to collect, or pool, in his legs. Weitz's heart worked hard to prevent the pooling and keep blood flowing normally through his body. This was a test to see if the astronaut's heart was getting lazy or deteriorating.

On Earth, the heart pumps blood "uphill" against the pull of gravity. In space the heart doesn't have to do this work, so it gets out of condition. If that vital pump got too weak it would not be able to keep blood flowing when the spacemen returned to Earth.

Weitz also rode a bicycle without wheels. This tested the ability of his muscles to do work. Floating around in weightlessness is like lying in bed as far

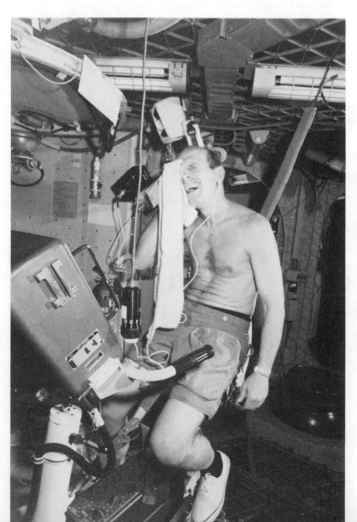

Astronaut Charles Conrad, Jr., commander of the Skylab mission, wipes perspiration from his face following an exercise session on the bicycle ergometer. The ergometer was the prime exerciser for the crewmen and measured how the strength of muscles was affected by a lack of gravity.

as muscles are concerned. Like the heart, they tend to get out of condition, so a close check must be maintained on their strength.

Having to pedal the bike as fast as he could in the hot workshop made Conrad irritable. Also, the astronauts kept floating off the bicycle because the holddown straps didn't work as planned. The first bike ride in space turned out to be a tough job.

Sun and Earth Watching

Tuesday, May 29, was the first time the astronauts settled down to doing the things they went into space to do. They turned on and used instruments in the Apollo telescope mount (ATM). The telescope mount, you remember, is a solar observatory with wings containing the only power-generating solar cells now working on Skylab. It was mounted on a truss structure over the MDA and the airlock module. However, the control panel was inside the MDA.

Without the Sun's light and heat no life could survive on Earth. Therefore, scientists want to learn all they can about how this star produces energy. This is difficult to do from the ground because Earth is surrounded by a thick envelope of moving air, clouds, and dust which acts like a veil. By putting telescopes and cameras on a spacecraft scientists would get the clearest look at the Sun in man's history. After the astronauts turned on the eight telescopes and instruments, scientists reported they were pleased with the way everything worked.

That night temperatures in the workshop reached 80 degrees, and engineers decided this was as low as they would get. Temperatures, therefore, would be 8 to 10 degrees higher than planned. But, with the low humidity, this was not uncomfortable for the astronauts. They slept in their living quarters for the first time that night.

Skylab instruments were designed to look down as well as up. Next day Conrad's crew did their first survey of Earth from space. For this, they used what is known as the Earth Resources Experiment Package (EREP)—cameras and other equipment which quickly scanned up to a hundred square miles of ground below Skylab. The information gathered could be used for mapping, detecting pollution, spotting storms and dangerous icebergs, surveying crops and forests, detecting crop and tree deseases, searching for mineral deposits and fish, and conservation of Earth's dwindling resources.

The spacecraft was rolled so that the EREP on the outside of the MDA faced Earth. Cameras and instruments began taking photographs over the coast of Oregon. They continued recording information on some twenty-five sites as Skylab traveled across parts of Nevada, Utah, Arizona, New Mexico, Texas, the Gulf of Mexico, Central America, Colombia, and Brazil. The pass ended as Skylab reached Brazil.

It was a successful survey, but it took its toll of electric power, in short supply since the accident that occurred when Skylab was launched. Flight director Neil Hutchinson explained the problem to reporters that evening: "Our power situation is such that we have been managing it very carefully. We have to always pay back to our batteries what we take out [by recharging them with sun energy converted to electricity by solar cells]. On the EREP pass today, we ran the batteries down to the lowest level."

One of these batteries could not be recharged. Another one had gone off on May 27, leaving 16 out of the 18 batteries working. With one wing on the workshop ripped away and the other unable to supply much power in a closed position, this left Skylab with a serious energy crisis. Flight directors canceled an EREP pass scheduled for next day. The crisis also affected medical experiments and observations of the Sun. Flight directors began planning a spacewalk to see if the astronauts could free the solar wing jammed closed by a metal strap from the heat shield.

Hanging by Your Toes

The strap that had riveted itself to the solar wing was about twenty-five feet from the exit door in the airlock module. Ground teams began to figure out how the astronauts could get over to the strap and hold themselves in place while they worked. They also had to determine which tools on the space station could do the job.

The situation in space was reconstructed in a huge water tank at the Marshall Space Flight Center in Alabama. An aluminum strap was fixed to the solar wing on a full-size model of Skylab. The spacewalk was worked out underwater because floating in water on Earth is the closest thing to being weightless in space. Astronauts Russell (Rusty) Schweickart and Edward G. Gibson wore spacesuits weighted so they would float at one level, rather than sink or rise. In this way, they tried different tools and different ways to free the now vitally needed solar wing.

By June 7, everything was set. One reporter called it "the high wire act of all time—without a net." Schweickart himself said: "It's like being on a trapeze . . . hanging by your toes. There are no normal handholds or footholds."

The astronauts themselves weren't too sure of their chances of success. "I think we've got a fifty-fifty chance of pulling it off," Kerwin commented.

"We're not quite as optimistic as you guys are," Conrad told the flight directors.

At 10:15 that morning, the airlock hatch was opened. It was dark outside, but lights lit up the immediate area. Conrad stepped out first. He got his feet firmly planted in foot restraints. Then Joe Kerwin handed him out 5-foot lengths of tubing. After Conrad assembled these into a 25-foot pole with strong cable cutters on the end Kerwin moved to the outside of the space station.

As Skylab orbited in the darkness within radio range of the tracking station at Carnarvon, Australia, Conrad reported: "The pole assembly went super slick.

Charles Conrad, Jr. (background) and Joseph P. Kerwin take part in the extravehicular activity to repair the damaged and partially deployed solar array wing. The wing was later fully deployed following the successful space walk.

We're peering around out here deciding how far Joe can get in the dark. I'm enjoying a lovely look at the Moon."

As they orbited beyond the Guam station and headed for the United States, Pete reported that he was watching the sunrise while upside down. When it became light enough to see, Conrad handed the pole to Kerwin who tied one end to a truss supporting the telescope mount. The next step was to work the other end of the pole so the jaws of the cutter gripped the aluminum strap. But Kerwin had a great deal of trouble doing this because loose wires got in his way and he could not hold his position. Kerwin's feet slipped out of the footholds and his legs went flailing in the air. His heart began to beat faster.

Skylab passed over the United States and out of radio range.

When contact was made again, the station was passing over South America. Conrad reported that the cutter's jaws finally gripped the strap. But the long umbilical cords that supplied him and Kerwin with oxygen, cooling water, and communication were tangled. Conrad, his heart pounding, untangled them.

There was no place to hold on and no footholds between the airlock and the solar wing. That's why the pole had been fixed in place on both ends. Conrad used it as a fireman's pole, moving hand-over-hand to the broken wing.

Exterior view of space station taken from CSM during final "fly around" inspection; cloudy Earth and horizon of Earth in background. One remaining solar panel is deployed on the right, relieving the tight electric budget.

He reached it as Skylab raced out of radio contact again. Thus far, 1 hour and 38 minutes had passed since the spacewalk began. While Skylab went around the world once more, Pete made sure the cutter jaws were in a good position to cut the strap. He signaled Kerwin, who actually made the cut by closing the cutter jaws with a rope. (The cutter worked just like the prunning tool used to cut high limbs on a tree.)

Kerwin actually hung by his toes over the edge of the workshop to get in the best position to cut the strap.

But the work was not over. Oil in the hinge used to swing out the wing was frozen solid by more than three weeks exposure to the intense cold of space. The wing would not swing out by itself. It had to be broken free.

For this, one end of a very strong rope had been attached to the telescope mount. Conrad hooked the other end of this on to the wing. Conrad got into position above the hinge. He squatted with the rope over his shoulder, and pushed against the spacecraft with his feet. As he did, the pressure applied to the rope broke the wing free.

Skylab came back into radio range off the west coast of the United States.

The tension in tracking stations all over the world eased when Conrad reported: "We've got the wing out and locked."

That night Skylab director Schneider reported: "We are getting electric power out of the wing." Controllers were maneuvering Skylab from the ground to get as much heat from the Sun as possible on the solar panels. "We are confident that if we get the proper heat in there, the panels will extend fully and we will get a great deal of electric power out of them," Schneider said. "The thing that has us so elated is that this will allow us to go off the tight electric power budget and to get very close to our original flight plan.

"I ought to take a few moments to tell you how personally proud I am of all the guys who did this," he continued. "This was a great team effort involving those in space and those on the ground. NASA has always felt that man can play an important role in space. We have shown that when you build things to be fixed, man, with that great computer he's got between his ears, is able to figure out a way to fix them. In the past we've proved that man can survive in space—can go up for brief periods and explore. Now we've shown that space is also a place where man can live and work—where he can achieve useful things."

Eating, Showering, and Hanging Up to Sleep

Things had quieted down after the successful 3-hour, 25-minute space walk. Paul Weitz was shaving and Joe Kerwin talked with astronaut Hank Hartsfield by radio.

"I have a message to read to you," said Hartsfield. "Everybody listening? 'On behalf of the American people, I congratulate and commend you in your successful effort to repair the world's first true space station. In the two weeks since you left the Earth, you have more than fulfilled the prophesy of your parting words: "We fix anything." All of us have new courage now that man can work in space . . . even as he does on Earth.' It is signed Richard Nixon, President of the United States."

The astronauts went to bed that night with the panels of solar cells fully extended. This nearly doubled the available power to about 7,000 watts. "Looks like we can turn on more lights and stop living like moles," Pete Conrad remarked.

With this new surge of electrical energy, the crew moved into a normal day-to-day routine that eventually turned a failing mission into an outstanding success.

The "Butter Cookie Monster"

A big part of normal Skylab routine involved cooking and eating. One astronaut was assigned as cook each day. He would check the menu, then

The three members of the prime crew dine on specially prepared food in the wardroom of the crew quarters. More than eighty food items were taste tested by the crew members. In zero gravity the crew used thigh and foot restraints to "anchor" themselves at the table.

arrange frozen, canned, and dehydrated food on three trays. Each tray contained eight holders, three of which could be heated for thawing and cooking items like packages of frozen lobster, filet mignon, and roast beef. All the astrocook had to do was turn on the heat and set a timer on the tray.

Magnets held knives and spoons to the tray. Slipperlike footholds and a lap bar enabled the astronauts to sit at a small table without floating away. Hot and cold water hoses at the table supplied drinking water and enabled them to prepare dehydrated food and beverages such as coffee, tea, cocoa, lime, lemon, grape, grapefruit, orange, cherry, apple, and instant breakfast drinks. These were packed in accordionlike plastic containers which the spacemen squeezed to get the liquid into their mouths.

More than seventy items were included in their meals. A typical breakfast chosen by Conrad consisted of scrambled eggs, sausage patties, strawberries, bread, jam, orange juice, and coffee. One of Kerwin's standard lunches comprised spaghetti, catsup, asparagus, bread, peaches, and grapefruit drink. One of Weitz's standard suppers included filet mignon, potato salad, strawberries, ice cream, and orange drink. Typical snacks included butter cookies, lemon pudding, mints, apricots and various drinks. Conrad became fond of eating two cans of butter cookies for an evening snack. He jokingly warned ground controllers not to disturb him while he was eating them. They, in turn, gave him the title of "Butter Cookie Monster."

Conrad once remarked that Skylab food was "like eating in the same restaurant every day." But he added that the food tasted much better and was more pleasant to eat than meals on his Gemini and Apollo flights. On Gemini missions in 1965 and 1966, he sucked gooey pastes out of plastic tubes and kept solid bite-sized chunks in his mouth until they became moist enough to stop tasting like wooden blocks. Going to the Moon, he and others spooned foods out of plastic bags to which hot water had been added.

It took many years of work by NASA engineers to develop food that astronauts could eat in weightlessness much the same way we eat on Earth. They made items like spaghetti and peas with a "sticky" consistency so they stayed on the spoon and didn't float away. They put thin coverings over the contents of cans. Astronauts could slice these open to get at the food, but the coverings kept the uneaten portion from drifting away. Even so, Conrad reported that "we were continually reaching out to get a ball of gravity or something else that got away."

Showers in Space

With the power shortage behind them, all three astronauts treated themselves to hot showers on June 8. About 6 P.M., Weitz stepped into a 42-inch wide hoop in the crews quarters. As he raised it, an attached shower curtain unfolded around him like an accordion. Weitz secured this to a lidlike fixture on the ceiling, turned on a spray nozzle, and strange things happened. Instead of showering down, the water broke up into a dense cloud of droplets that floated around, splattering when they hit him or the shower curtain.

It was a good thing that the water floated around as it did instead of going down the drain. The space station carried just 6,000 pounds of water, so each

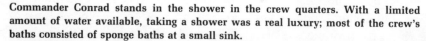

Commander Conrad stands in the shower in the crew quarters. With a limited amount of water available, taking a shower was a real luxury; most of the crew's baths consisted of sponge baths at a small sink.

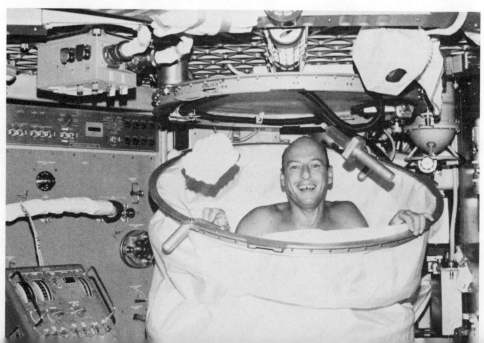

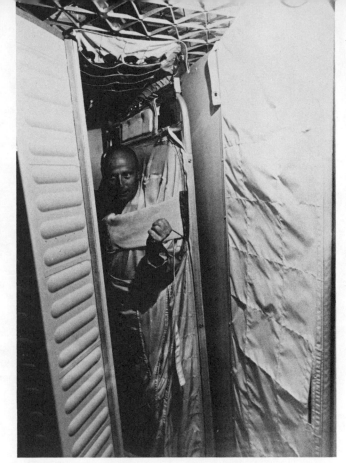

Charles Conrad, Jr. demonstrates how to go to bed in his upright sleep restraint, which prevents drifting around because of weightlessness. The crew's sleeping quarters are in the lower level of the two-level Orbital Workshop.

astronaut was allowed only three quarts for one shower a week. As long as the water kept drifting around in the enclosure, this was enough. Weitz wet himself down, turned off the spray, and washed himself with a soap-impregnated towel. Then he used the rest of the water to rinse off.

Now came the hard part—getting water out of the shower enclosure. Weitz turned on a pump which drew some of it away. Then he used what looked like a small vacuum cleaner to remove the drops clinging to the curtain. "Fifteen minutes to clean myself and forty-five more to clean up the mess," Weitz thought. "It sure takes a lot of time."

On non-shower days, the spacemen cleaned themselves at a small sink, using soapy washcloths and towels.

Hanging Up to Sleep

Conrad, Kerwin, and Weitz slept like bats, that is, hanging from the ceiling. Each had his own private room—about the size of a large closet. Hanging from the ceiling and tied to the floor was a type of sleeping bag. The astronaut enclosed himself in the bag to keep from floating away as he slept. Since "up" and "down" as we know them on Earth do not exist in weightlessness, there was no need to lie down. You could be comfortable sleeping or floating in any position. Most Skylabers slept with their heads toward the ceiling, but Alan Bean, commander of the second mission, slept with his head toward the floor. It's all the same when the pull of gravity can't be felt.

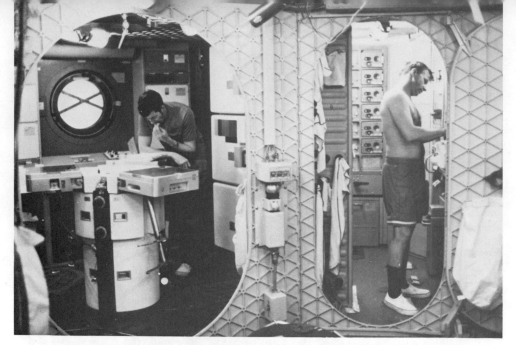

Two members of the prime crew are shown in the Orbital Workshop. Joseph P. Kerwin is in the wardroom studying the wiring plan, and astronaut Paul J. Weitz is preparing for personal hygiene in the space station's bathroom.

Weitz didn't like sleeping in his cramped apartment, so he hung himself up to sleep where he wasn't so close to a wall. "Every night P. J. disappears, and we don't see him until the next morning," Conrad reported to the ground. "We call him the night wanderer. Joe and I run about six and a half hours' sleep at night, but Paul gets along on less.

"It's very quiet up here," Conrad continued. "There's no vibration at all. The important thing is that all of us can do all the tasks, and we really don't have too much trouble doing them. We've adapted very well. If we're reading, we free-float and wind up wherever we wind up—on the ceiling, on the floor, ricocheting off the walls, even in the corner. And it doesn't seem to bother us."

After a while, he said, everything looks perfectly natural, even standing on the ceiling.

"I'm really convinced that the first day back, we're going to leap out of bed and land right on our heads."

Going to the Toilet

Imagine going to the toilet in a place where everything floats around freely instead of falling down into a receptacle. Imagine further that you must collect a sample of your body wastes each time you go to the toilet. This is what Skylab astronauts had to do.

In the bathroom next to their sleeping quarters, a toilet seat was mounted above what NASA calls three "urine drawers." Foot straps were supposed to hold the astronaut in place while he used the drawers, or seat. But Conrad's crew found they couldn't get their feet under the straps while wearing cleated shoes. Instead, they grabbed handholds next to the toilet seat for urinating.

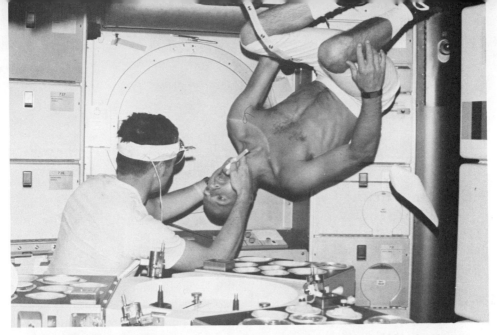

Charles Conrad, Jr., crew commander, gets a physical checkup from Dr. Kerwin. Note floating paper, which along with the upside down crew commander indicates the weightlessness of the Skylab space station environment.

These handholds and a lap belt held them in place on the seat. The belt didn't work too well, either. They couldn't tighten it enough to hold them securely on the seat.

An air blower provided a substitute for gravity and drew the waste into a container. When a crewman used this flushing system at night, it made so much noise it woke the other astronauts. Conrad, Kerwin, and Weitz did the best they could without complaining too much.

Each time an astronaut used the toilet, he removed the container with his feces, weighed it, dried it in an oven-type device, then stored it for return to Earth. Once a day, each spaceman measured the urine in his drawer and took a sample of it. They stored the samples in a freezer for return to Earth along with the feces, blood samples, and any vomit. A filter system removed all smells from toilet and storage areas. Handling waste required a lot of time and effort, but doctors regarded it as absolutely necessary to determine if man could adapt to living in space for long periods.

During flights to the Moon and in Earth orbit, both American and Russian astronauts lost weight, their muscles shrank, and minerals like calcium left their bones. These effects disappeared after a few days back on Earth. But medical experts feared that on long missions bones and muscles would become dangerously weak.

The big question at the beginning of the Skylab program was whether these effects stopped after so many days in space or continued to the point where astronauts became too sick or weak to do their jobs. To answer this question, the astronauts themselves and the samples they returned would be carefully checked to see how much of what minerals, hormones, and blood elements left their bodies. The results would be used to determine whether or not it would be safe to send a second crew up for 56 days.

A Change of Heart?

Blood and waste samples were, in fact, only part of a broad series of medical experiments to determine if man could survive for long periods in space. When man first went into space in 1961, some doctors feared that all sorts of horrible things would happen.

"People were concerned that astronauts would choke if they tried to eat because the food wouldn't move down into their stomachs without gravity," recalled Richard S. Johnston, a NASA official. As men flew higher and faster, some feared that a spaceman's heart would beat so rapidly and his blood pressure rise so high that it would endanger his life. Others worried that the human body could not stand the stress of accelerating to 17,000 miles an hour, then decelerating from such speeds and plunging back into Earth's atmosphere. These fears and worries proved false. However, enough changes were found in the bodies of cosmonauts and astronauts so that many questions about the ability of men to survive and work efficiently in space needed to be answered.

Since Skylab involved more men staying up for longer times in larger spacecraft than before, this was the chance to get some answers. In addition to checking changes in bones and muscles, experiments tested the effects of space-

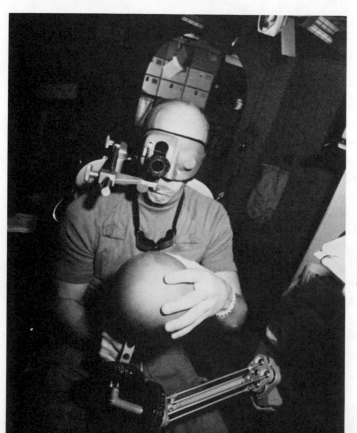

Astronaut Conrad checks the body responses to the effect of reduced gravity and Corealis forces. The litter chair in which he is seated can be rotated by a motor at its base. It can also be titled forward, backward or to either side.

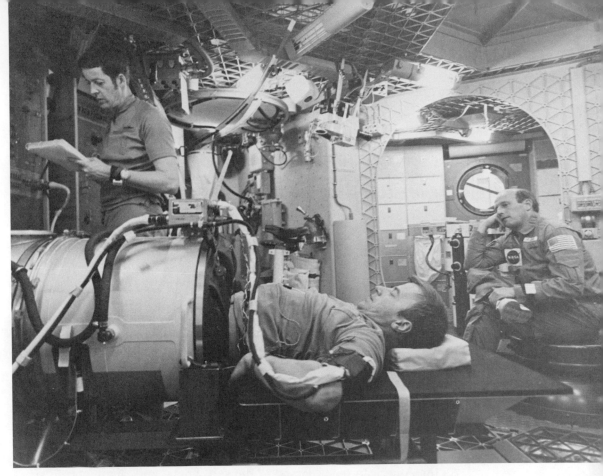

Commander Conrad (right) watches as Dr. Kerwin checks Paul J. Weitz in the lower body negative pressure machine, which places stress on his heart and blood vessels. This provides information concerning heart function during space flight.

flight on heart action, blood cells, body fluids, changes in the body's natural resistance to disease, and variations in chromosomes and genes. Regularly scheduled tests checked the astronauts' nervous systems, especially their balance, ability to stay oriented while floating in space, and resistance to motion sickness. Joe Kerwin wore an electronic sleeping cap to measure the effects of spaceflight on his sleep and dreaming. Time and motion studies were done to compare how men performed tasks in space and on the ground.

One of the most important sets of experiments involved studies of the heart and blood vessels. These were conducted while astronauts pedaled the wheelless bike or laid in the drumlike lower body negative pressure device. One day while Joe Kerwin was in this device, his heart suddenly speeded up and his blood pressure dropped. This meant that his heart was having difficulty pumping out blood collecting in his legs. Ground controllers stopped the test. On June 11, Paul Weitz experienced the same problem.

This indicated that the astronauts' hearts were changing. Doctors expected that their hearts would not be able to do the same amount of work that they did on Earth. This was part of adapting to space, where the heart does not have to work against the pull of gravity. The big question was whether heart changes occurred which harmed the astronauts in space, or made it difficult for them to survive when they returned to Earth. If either of these things happened, long space flights might not be possible.

Russian Record Is Beaten

On June 18, everyone felt that things would turn out in favor of long space journeys. At 2:22 A.M., Central Daylight Time, as Skylab orbited over Nigeria, Africa, Conrad, Kerwin, and Weitz broke the record for the longest spaceflight in history. This had been established on June 30, 1971, when cosmonauts Georgi Dobrovolski, Vladislav Volkov, and Viktor Patsayev completed 570 hours and 22 minutes (almost 24 days) in space. When this record was broken, the Skylab space station was on its five hundredth revolution around Earth and its crew had traveled 9,500,000 miles.

These three cosmonauts were the ones who died as they returned to Earth. Air rushed out an exhaust valve accidently opened by the firing of the braking rockets that brought them down. Before the confused spacemen could close it—in a matter of ten seconds—too much air had leaked out for them to remain alive.

As the American team passed 23 days, 18 hours, and 22 minutes in space, Conrad asked chief astronaut Donald "Deke" Slayton to relay his crew's respects to Russia's cosmonauts and wish them good luck for the future.

Next day, the Americans received a return message from Vladimir Shatalov, a veteran of three space flights and chief cosmonaut: "We sincerely congratulate the courageous crew of Skylab astronauts on your achievements in conquering outer space. Wishing you successful completion of your program and safe return to our beautiful blue planet Earth. On behalf of the team of Soviet cosmonauts."

Chapter 5

Splashdown

"I'll tell you when he hits it, Joe," Paul Weitz said. "There it goes. Boy, is he hitting it. Holy cats!"

Weitz was standing in the open door of the airlock module watching Conrad, who was clinging to the outside of the space station. Pete held the telescope mount with one hand and swung a ballpeen hammer with the other.

They had tried every way but this one to loosen the stuck regulator. It was supposed to charge a battery by feeding it current from solar cells on the telescope mount panels, but it had not been working since the beginning of the mission.

Conrad whacked it again.

Joe Kerwin was in the docking adapter listening to Weitz, watching gauges and working the controls of the battery charger. Suddenly, a needle jumped on one of the gauges.

"I turned the charger on, and I'm getting a lot of amps plus [current] on the battery," Kerwin told the ground.

"It worked! You've done it again," came the reply from the control center.

Skylab was orbiting over the United States on June 16 as Conrad and Weitz took the mission's second space walk. After fixing the battery regulator, Conrad placed film in six of the eight experiments located on the telescope mount. Since the human eye cannot look directly at the Sun without being damaged, cameras automatically took photographs of it every day, recording things that cannot be

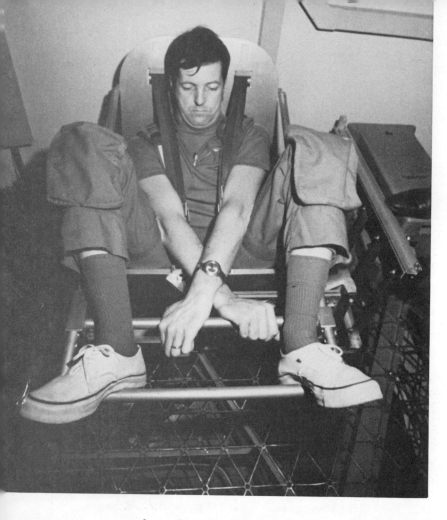

Science pilot Joe Kerwin demonstrates the Body Mass Measurement experiment. The device, a pivot-mounted chair, will be used for daily determination of the crewmen's weight.

seen from the Earth's surface. Scientists all over the world were anxious to view this film.

Conrad worked his way to the hub of the windmill-like solar panels, about fifteen feet above the airlock module. He slipped the toes of each boot under restraining bars. Then he locked heel clips on his boots into special fittings. He looked toward the airlock, where Weitz was making sure the thick cable supplying him with air and radio contact did not get tangled up.

Conrad opened a door on the back of an instrument and removed a film magazine. Weitz had swung out a boom, on the end of which was a "film tree." Canisters of fresh film hung from it like huge ornaments on a leaveless, metal Christmas tree. Pete removed a fresh load of film from the tree and put the used magazine in its place. Then he loaded the new film into the instrument. He did this twice at one location and four times at another until all the cameras were reloaded. The smallest magazine was about the size of a large, thick dictionary. The largest was about as big as a typewriter case and weighed sixty pounds.

Conrad had no trouble with the film changing. He then worked his way to another part of the telescope mount. With a small brush he reached around and swept away a piece of debris which partly blocked the view of one instrument.

The veteran Navy test pilot sounded like he was having fun on the spacewalk.

"Wheee, is that a pretty sunset! [I'm looking at it] upside down. Wheee!" Ground controllers heard him say.

Then Conrad tackled what turned out to be the toughest job of the spacewalk. He had to attach a square of the material used to make the parasol on to one of the tubular struts supporting the telescope mount. The sample would be brought back to Earth by the next Skylab crew so engineers could tell how well the material was holding up in space.

"The gosh darn stuff is hard to handle out here," Pete radioed. "It doesn't want to do what I want it to do."

His heart beat as fast as a hundred and fifty times a minute as he struggled with the sample of cloth. This is more than twice the normal beat of about seventy times a minute. Finally Pete succeeded in getting it in place. With the help of handrails, he worked his way back to the airlock. The two astronauts stepped inside and closed the door. All the tasks had been completed. They had used up only one and a half hours of the three hours allotted for the spacewalk.

"We had a very good day today," flight director Milt Windler told reporters that afternoon.

Garbage Problem

The next two days—June 20 and 21—the astronauts spent packing and closing up their spacehouse. They moved film, medical samples, records, and personal gear into their Apollo ferry ship. They shut down some equipment and set up other instruments so data could be taken while the station was unmanned. And then they cleaned up the place.

This job gave them a problem on the morning of the twenty-first. Conrad put a large canister containing charcoal from an air filter and four used gloves into a disposal bag. In the center of the crew's quarters stood a large round "trash dump." You opened a lid, put the trash in, then closed and locked it. Then you stepped on a lever which opened a lower or outer door and dumped the trash into a tank below the floor.

When Pete put the bag in the airlock, the canister hung up on the lower door in such a way that he couldn't push it down or pull it back out. Conrad tried to open the door with the dumping lever, but the lever stuck. A jammed trash lock could create real difficulties. Three men in space throw away a lot of dirty clothing, air filters, food cans, urine bags, tissues, washcloths, towels, and other items. Without a place to get rid of it, this trash could create health and space problems, especially for the crews scheduled to spend 56 days in space.

Everyone on the ground began working on the problem. However, about a half hour after it occurred, Conrad got on the radio. "By judicious application of muscle, we managed to get it up and free," he announced. "The trash airlock is operative one more time."

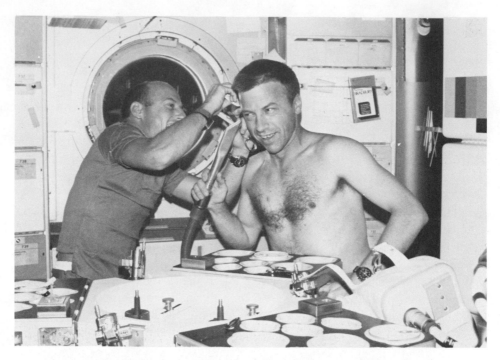

Astronaut Conrad trims the hair of Astronaut Weitz during their mission. They are in the wardroom of the Orbital Workshop. Weitz is holding a vacuum hose in his right hand to catch loose, floating hairs.

"Boy, is that ever good news. You can hear the sighs of relief down here," said a ground controller.

"They're nothing compared to the sighs up here, man," Conrad told him.

Bye-bye Skylab

Ground communicators wakened Conrad, Kerwin, and Weitz at 1:08 A.M. on June 21. They worked to close down the station until 2:30 that afternoon, then went to bed. They set their alarms to awaken again at 7:30 that evening in order to prepare for an early return to Earth next day. Their first task was to turn on everything in the Apollo spacecraft and check all its systems to be sure that no problems had developed during 28 days in space.

As they did this, ground controllers keeping check on the station discovered that temperatures were rising in the refrigeration system used to chill drinking water and to keep frozen foods, urine and blood samples at the proper temperature. Joe Kerwin went back into the station to see that all the controls had been properly set. They were, but the refrigeration system kept warming up. It was a bad time to have a problem.

On the ground, engineers thought that lines in the system might have frozen. The astronauts were busy with other things, so ground teams fired maneuvering rockets on the station by remote control. They tipped the station nose down to

Exterior view of space station taken from Command Service Module during a "fly around" inspection; dark sky background, looking toward docking port in the Multiple Docking Adaptor. Solar panel is on the left; two antennas protrude downward.

bring the area where the problem existed into the Sun for about an hour. Warming up the lines, they thought, would melt the frozen liquid that was clogging them.

Meanwhile the astronauts continued with their routine, putting on their spacesuits and getting ready to come home. Flight directors discussed whether there was anything the crew could do to help solve the refrigeration problem. They decided that there wasn't. They might as well come down. However, when it came time to undock the Apollo ship, the station hadn't settled down into the proper position after the maneuvers by commands from the ground.

It was 3:34 in the morning, and the astronauts watched the minutes slowly tick away. Finally, at 3:54, this message came to them from the tracking station at Goldstone, California: "You're GO for undocking."

With a sigh of relief, Conrad took the controls of the Apollo spacecraft and began to back away from the spacehouse.

"Okay, we're free," he said. "We're moving away at about three miles per hour. Bye-bye, Skylab."

A Flaming Return

Conrad, Kerwin, and Weitz flew around their home in space for one last check. Then at 4:24 A.M., while orbiting over the Indian Ocean, they fired small jets on the spacecraft to drop them into a lower orbit.

At 5:05 the astronauts fired the large main engine on the rear of their spacecraft, slowing it down and dropping it into an orbit with a low point 104

An overhead view of Skylab space station, silhouetted against black sky, taken from CSM during "fly around" inspection. Conrad, Kerwin, and Weitz flew around for one last check on their home in space before heading for their homes on Earth.

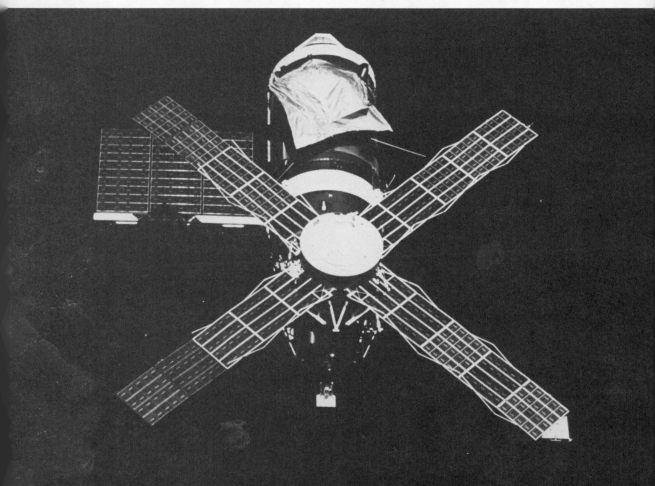

miles above Earth. This was over the Philippine Sea in the western Pacific. The spacecraft circled the Earth almost twice more. Then, as it came over Thailand again, the big engine fired for a full seven minutes. This braking action both dropped the craft out of orbit and aimed it at a landing spot. The astronauts started down on a long arcing path that would end in the Pacific Ocean 830 miles southwest of San Diego, California.

The rear part of the spacecraft, containing the main engine and other systems was called the service module. It was no longer needed, so the astronauts separated it from the cone-shaped front part, or command module. With the crew strapped securely in their couches, the blunt, rounded end of the command module slammed into the Earth's atmosphere at an altitude of 40,000 feet. It was like riding a meteor. The rounded end of the craft glowed like a red-hot coal. The astronauts saw the fiery red and purple light out the windows. They watched blazing pieces of the heat shield fly past. It was both thrilling and frightening.

The astronauts stayed comfortable inside their well-insulated spacecraft, even though the heat shield reached temperatures of thousands of degrees. About ten minutes after the spacecraft plunged into atmosphere, two small parachutes spilled out of its narrow end. They slowed the ship enough to allow three large main chutes to come out and open up. The astronauts could see the big orange-and-white-stripped canopies fill with air to cushion their drop into the Pacific Ocean. They splashed down at 8:49 A.M.—28 days and 49 minutes after their blast-off. Cheers went up from television watchers all over the world. At the Skylab Mission Control Center, cigars were lit, hands shaken, and members of the ground teams smacked each other on the back.

Kerwin Gets Sick

"Splashdown was not overly hard," Kerwin reported. Where they landed, it was a warm 67 degrees. Twelve-mile-an-hour winds stirred up waves one or two feet high, but long swells raised and dropped the spacecraft five or six feet every few seconds.

Kerwin took everyone's pulse and found the astronauts close to normal. "I then hustled down into the lower equipment bay and got everybody a strawberry drink," he said later. "I chug-a-lugged my drink, and immediately knew I had made a mistake. . . . I got seasick," confessed the Navy doctor.

All three astronauts felt lightheaded—as if the spacecraft were rotating around them. Their arms and legs felt heavy. This was the result of coming back into Earth's gravity after floating weightless for 28 days. But Kerwin felt worse than this.

Superimposed on the lightheadedness and heaviness, he said, "was this awful feeling that the world was about to swallow you up. It stayed with me through the entire day. Naturally you feel better after you throw up, and I managed to accomplish that feat."

Kerwin had not thrown up yet, when the 912-foot-long aircraft carrier USS *Ticonderoga* maneuvered alongside the tiny, bobbing spacecraft. After missions to the Moon, returning astronauts left their spacecraft and got into a rubber life raft. A helicopter then lifted them from the raft and carried them to the recovery ship. On the Skylab mission, the aircraft carrier crew hoisted the spacecraft directly onto the deck with the astronauts inside. Doctors wanted the space voyagers to remain as inactive as possible until they started their medical tests. This would enable doctors to better determine the effect of 28 days of weightlessness and how the men were adapting to the return to Earth.

Thirty-nine minutes after their command module hit the water, the astronauts were safely on the deck of the *Ticonderoga*. The ship's band played "Anchors Aweigh" as the astronauts stepped out of their craft one by one and saluted the flag. All looked and felt a bit dizzy and shaky. But they managed a proud smile as they walked to the special medical laboratory set up to receive them.

Back at Skylab Control at the Johnson Space Center, Dr. James Fletcher, overall head of NASA, told a press conference: "We're all pleased and excited that the Skylab crew has arrived safely. It was a spectacular recovery, one of the smoothest. Essentially all the objectives of the mission have been completed. None of us really dreamed that this could be done at the time the meteoroid shield failed. The mission has exceeded our wildest expectations."

William Schneider, Skylab's director, described the crew's accomplishments in detail. "Despite the major problems we had, we achieved over eighty percent of our objectives with the telescope mount," he said. "We obtained over thirty thousand pictures. In looking at these, scientists have already seen things they never suspected before on the Sun.

"Although Earth resources experiments were limited because of the power problem, we were able to look at one hundred eighty-two target sites in thirty-one states and nine countries. [A total of 16,765 photographs were made with three cameras, and 42,500 feet of tape were used by other EREP instruments to collect data.] We did ninety percent of the planned medical experiments."

"Space Is Beautiful"

The astronauts rapidly recovered from their lightheadedness and the feeling that their bodies were heavy. By the second day after splashdown, even Kerwin felt fine.

"If this is the worst that space can do, we're up there to stay," he told everyone. "Space is kind to people. Space is beautiful."

"In flight," Kerwin added later, "it was a continuous and pleasant surprise to me to find out how easy it was to live in zero gravity, and how good you felt."

On Sunday, June 24, Conrad, Kerwin and Weitz flew to San Clemente, California, to meet President Nixon and the leader of the Soviet Union, Leonid I.

At the former California White House, Soviet Communist party leader Leonid I. Brezhnev and ex-President Nixon examine plaques presented by Skylab astronauts Charles Conrad (center), Paul Weitz (left), and Joseph Kerwin.

Brezhnev. They gave the Russian leader a Skylab patch and a pocketknife carried on the mission. He called their flight a "great achievement" and invited them to visit the Soviet Union.

Their hearts beat faster than normal when they first returned to Earth. But the beats soon slowed to normal. The calcium lost from their bodies was not enough to affect the strength of their bones. Conrad had lost four pounds, Kerwin six pounds, and Weitz almost eight pounds. Most of this loss came from their legs. Their thighs and calves shrank as much as a inch. But this was expected and no one worried about it. Conrad recommended that the next crew do additional exercises to keep their leg muscles in shape.

The Skylab commander said that exercise was extremely important in keeping his crew in physical shape while in space and enabling them to readapt to conditions on Earth.

"The exercises that we did contributed significantly to our well-being up there," he said. "Without them, I'm convinced they would have carried us out of the spacecraft rather than us getting up under our own steam and walking." Conrad recommended, and doctors agreed, that the next crew should increase their exercise time from a half hour to an hour, or an hour and a half each day.

With the addition of more exercise, Conrad said he thought the next crew could easily spend 56 days in space. Dr. Royce Hawkins, in charge of examining Skylab crews, agreed.

Kenneth S. Kleinknecht, manager of the Skylab program at the Johnson Space Center, added his assent. "We have no restrictions that would keep us from flying the next mission," he said. "The launch date will be on the twenty-eighth of July."

Chapter

*Second
Mission
Blast-off*

It was foggy the morning of July 28 as the second Skylab crew waited atop the Saturn IB rocket at Cape Canaveral. Sitting in the commander's couch on the left side of the command module was Navy Captain Alan Bean, a forty-one-year-old test pilot from Texas who had been the fourth man to walk on the Moon. He had spent 31 hours on the lunar surface with Pete Conrad during the second Moon landing mission in November 1969. In the middle couch sat scientist-astronaut Owen K. Garriott, almost forty-three, a native of Oklahoma, an electrical engineer and a physicist. Owen was thinking about the minnows, mice, bugs, and two spiders named Arabella and Anita that he had brought aboard the spacecraft. Next to him sat Jack Lousma, thirty-seven, a handsome Marine Corps pilot who was so relaxed he fell asleep during the countdown. It was the first spaceflight for Garriott and Lousma.

The Saturn IB rocket that the astronaut crews rode into space was smaller than the huge Saturn V Moon rocket that boosted the heavier space station into orbit. Saturn V stood 363 feet tall, Saturn IB measured 224 feet. The Moon rocket weighed 6.2 million pounds and its first stage developed a thrust of 7.5 million pounds. Its smaller cousin weighed 1.3 million pounds and boasted a thrust of 1.6 million pounds.

As the minutes ticked away at the Florida launch pad, the Skylab space station orbited over the Atlantic Ocean, south of Newfoundland. It was making its 1,085th revolution around Earth since launching on May 14. Temperatures in

The second Skylab crew was launched from Pad B, Launch Complex 39, Cape Canaveral Space Center, Florida, at 6:10 a.m. July 28, 1973.

the troublesome refrigerator system had settled down and the food was staying cool enough not to spoil. NASA had decided to extend this second mission from 56 to 59 days. This would allow the astronauts to come down much closer to the United States at the end of their journey, so that medical checks could begin at once. Skylab's new flight path would bring them down in the Pacific 340 miles from San Diego instead of 1,200 miles away, as originally planned.

It was now 6:10 A.M. Central Standard Time. Three miles from the rocket, at the launch control center, a NASA official counted the final seconds: "T minus 10, 9, 8, 7, 6, 5, 4, 3 . . ." At this point eight engines at the base of the rocket flared to life. In the next three seconds they built up a thrust of 1.6 million pounds. At the count of zero, when full thrust was reached, the rocket began to move upward. Cheers went up from spectators and flight controllers as the twenty-story gleaming white rocket rose out of the brilliant orange flame and white smoke.

It rose slowly at first, then picked up speed. In 36 seconds the rocket reached a height of 3,700 feet and a speed of 975 miles an hour.

This was the emblem for the second manned Skylab mission. The patch symbolizes the main objectives of the flight—the study of man himself in the zero-gravity environment of space, the study of the Sun, and the development of techniques for surveying the Earth's resources.

"It's got a pretty noise to it right now," Alan Bean said above the roar.

"That thing really shook, rattled, and rolled getting off the pad," exclaimed Lousma. "Now it's a great feeling of motion. I really feel like it's moving out."

After 2 minutes and 16 seconds the first-stage engines began to cut off. Five seconds later, the 80-foot-long first stage dropped away and fell end over end into the ocean. Then the 58-foot-long second stage flamed to life. It gave the spacecraft a 225,000-pound additional push. The astronauts reached a height of 41 miles and a speed of 5,200 mph only 2 minutes and 45 seconds after launch.

"Hey, we'd like to try that lift-off again. That was great," said Lousma.

Eight minutes after they left the launch pad, the spacemen had traveled 745 miles from Cape Canaveral. Their spacecraft was 100 miles high and moving at 13,000 mph.

At 10 minutes after lift-off, the second-stage engine cut off and they glided into orbit. The astronauts were now traveling 17,500 mph. The second stage separated and the spacecraft continued on alone.

A Sparkling Leak

Now began an eight-hour chase to catch the space station. The astronauts began a series of maneuvers with their main engine and smaller maneuvering jets which would bring the two vehicles together. Thus far, the launch had been one of the smoothest in the history of the space program, and ground teams began to relax.

But not for long. Three hours into the flight, controllers noticed that the level of fuel in one of the four sets of small maneuvering rockets had started to drop. The sets are equally spaced around the outside of the service module. Each has four rockets or thrusters. They are used for small maneuvers in space, and to bring the spacecraft back to Earth if the main engine fails.

"We got some sort of sparklers going by the right-hand window," Bean reported.

"It looks like we've been driving through a snowstorm real fast," Lousma commented.

As fuel leaked out, it froze immediately into particles like large snowflakes. When sunlight hit them, they sparkled.

Engineers on the ground told the astronauts to shut off the leaky set of four thrusters, called a "quad." They could maneuver all right with the remaining three quads. When the astronauts shut off the thrusters, the leak stopped. Flight directors decided they could go ahead with the mission.

At 12:27 P.M., Bean spotted the space station about 83 miles ahead. "We can see the solar panels on the telescope mount real well," he told the ground.

The Skylab space station cluster in Earth orbit is photographed against the horizon of the Earth. The area below is clouds over water, the area above a black sky background showing how the space station looked as Bean's crew overtook it.

Two more maneuvers were made, and the Apollo craft caught up with the station at 2 P.M. that afternoon. Bean began to fly around the 85-foot-long spacehouse to check it out. As he did, exhaust gases from the quad thrusters hit the parasol and made it flap around.

"Funny ole parasol is really blowing in the breeze," Lousma told the ground. "Al [Bean] thinks he'll knock off the fly-around to avoid blowing that parasol away."

"We think that would be the thing to do," astronaut Dick Truly told them from the Houston control center.

Bean flew around to the front of the house. He eased the spacecraft into its "garage" as the two vehicles orbited over the Atlantic Ocean off South Africa.

"We're docked," he reported when radio contact was established through a station in Australia. "That went real well. "We're having a snack, then we're going to get after it."

Astronauts Get Spacesick

The crew ate peanut butter cookies, peaches, vanilla wafers and drank orangeade. At 3:30, they started to open the tunnel between the two craft. They prepared to eat supper and sleep in the house that night. But at this point, another problem overtook them. Jack Lousma got sick.

At the time the spacecraft went into orbit, he felt queasy. Later, he threw up his lunch. Lousma took a motion-sickness pill, and he began to feel better. But later, as he worked in the station, the Marine felt nauseated when he moved his head rapidly. Instead of the regular evening meal, Lousma ate butterscotch pudding, an instant breakfast, and some orange drink.

Bean and Garriott didn't feel well either. They moved slowly to keep from becoming dizzy, and felt unpleasant sensations in their stomachs. It was like being seasick or carsick.

Both the astronauts and NASA doctors thought the sickness would pass quickly, but it did not. Next day, they could not eat breakfast, and Lousma vomitted again—twice. The astronauts worked so slowly that they fell a half day behind in getting the space station ready for living and experiments. Doctors ordered them to take motion-sickness pills, but the medicine didn't help much.

"There's a desire to take a break for an hour or two and get in the bunk," Bean said.

Doctors decided to give the crew a few hours off to rest and settle their stomachs. But as the three were trying to get comfortable an alarm went off, indicating an electrical problem in the Apollo spacecraft. Everyone began "trouble-shooting" the problem.

It was discovered that an electrical short circuit had ruined two biology experiments, killing six pocket mice and a swarm of vinegar gnats. These were carried in special cages in the spacecraft and were to be checked automatically by instruments to see how a space journey affected their body rhythms. All animals and plants have a mysterious internal timing device, or biological clock, which keeps them in harmony with their surroundings. This clock adjusts the vital processes in the body so that they do the most good during different parts of the twenty-four-hour day-night cycle. Scientists want to know if these biological clocks can be changed by outside factors, such as changes in gravity. If spaceflight changes body rhythms of mice and gnats, then it might also effect the health and well-being of astronauts.

Automatic devices kept the animals alive and sent the information back to Earth without the astronauts having to do anything. However, once power to the cages was shut off, the devices failed and the animals died.

Situation Gets Worse

Working on this problem didn't make the astronauts feel any better. Doctors ordered them to take more motion-sickness pills and to do head-movement exercises. The medical men hoped that the head movements would enable those

Skylab astronauts: Dr. Owen K. Garriott, science pilot; Jack R. Lousma, pilot; and Alan L. Bean, commander. All three astronauts suffered from motion sickness and fell a day behind in their activities.

parts of the inner ear that control balance and dizzines to adjust to weightlessness faster. Dr. Royce Hawkins explained that it wasn't unusual for astronauts to get sick when they first went into space.

"We've seen the same thing on practically every flight we've flown," he said. "The unusual thing this time is that all three got sick, and the sickness is so severe."

Dr. Hawkins explained that some astronauts naturally get motion sickness easier than others. The amount of movement that this crew made to get the station ready was enough to make all three members sick. Because of the meteorite shield problem, Conrad's crew spent much more time in the small area of the spacecraft before floating around in the much larger space station. They had more time to adjust to weightlessness before they had to move around so much. Also, before their flight, the first crew did more head exercises and acrobatic flying to get accustomed to motion which could make them sick.

Bean's crew felt a little better on Monday, July 30, but they still moved slowly and got further behind in their work. A spacewalk planned for Tuesday had to be postponed. The astronauts missed an important flare, or explosion, on the Sun because some of the telescope instruments ran out of film that should have been replaced on the spacewalk. Medical experiments were not conducted. To make matters worse, the system which removed moisture from the air sprang a leak and the station became very humid. Temperatures for the most part remained in the high 70s. Things might cool off if the astronauts put up another sunshade, as they planned, but the three were too ill.

If the sickness continued, would they have to be brought back to Earth? reporters asked Dr. Hawkins.

"There is a limit to how long you can keep them on motion-sickness pills," he admitted. "If they are not able to work after that time, yes, we'd have to consider an alteration in the mission."

"We Have a Mission"

Lousma was able to keep all his food down on Monday and again on Tuesday. He reported that he felt better. Garriott also said he felt "very fine." Bean complained that "at mealtime it is a little tough. As long as you can get through the meals, you are great between them."

Tuesday morning, they worked with two minnows and two spiders that Garriott brought aboard. Scientists wanted to see how absence of gravity affects the way fish swim and the way spiders spin their webs. The spider experiment was suggested by a seventeen-year-old student, Judith Miles, from Lexington, Massachusetts. Garriott reported that the fish seemed disoriented and confused in space—like the astronauts themselves. He then set up an automatic device which took movies of the spider trying to build its web. (How the fish and the spiders fared is explained in Chapter 8.)

The astronauts ate better and took no pills during the day. They even exercised on the wheelless bike. Al Bean jogged around the walls. "I think we have a crew and we have a mission," NASA doctor George Armstrong told everyone that night. "They're still moving a little slow, but they're not feeling bad and they're functioning."

The next day, August 1, doctors suggested that the crew eat four to six snacks, instead of three big meals. When they tried this, Al Bean reported that it was the first time they felt good after a meal. They began medical experiments on each other. All three moved their heads rapidly without getting sick. Plans were made for an earth resources survey on Friday and for the spacewalk on Saturday. On a live television show beamed to Earth, Al Bean showed everyone how you could eat upside down in weightlessness. The astronauts joked and played music.

NASA provided the astronauts with a special line on which they could talk to their families without anyone else listening. That evening Bean called his wife, Sue, and his children—a seventeen-year-old son and a daughter, ten. Later, Owen Garriott called his wife and four children, ages eighteen, sixteen, twelve, and seven. Jack Lousma also used the special phone to contact his wife and three children—nine, six, and five years old.

Emergency!

After what they called "the best day yet," it looked as though things would start to go smoothly for the astronauts. But things didn't. At about 5:50 next morning, an alarm woke them up. It came from the system of four quads, or sixteen small thrusters, spaced around the Apollo service module. There was a drop in fuel pressure in the quad opposite the one that leaked on the first day of the flight. Engineers suspected a second leak now. "When we looked out the window this morning," Bean said, "we noticed lots of sparklers."

This was a grave situation. The leaking thrusters could be shut off, and they would not affect the mission while the astronauts were in orbit. However, when they tried to come back to Earth, this would be another matter.

The big main engine brakes the 17,500 mph speed enough for the astronauts to begin a controlled fall back to Earth. Should the main engine fail, the maneuvering quads can be used as an emergency system to get them down. If the main engine remained okay, and two of the four quads worked, everything would be all right. What worried NASA officials, was the possibility that the two leaks on either side were related. They feared that something had gone wrong with the system which supplies fuel to all the quads. This meant that all four might fail before the end of the mission.

The leaking substance was nitrogen tetroxide. It is combined with another chemical called hydrazine to form the explosive mixture that powers the small thrusters. The nitrogen tetroxide could cause an explosion if enough of the leaking material collected inside the spacecraft.

Top officials held quick meetings and decided they had three choices. First, they could bring the astronauts down right away. Second, a rescue ship could be sent up to get them. This would take at least a month, but the astronauts could carry out many experiments in the meantime. Third, they could wait and hope that the leaks were unrelated and that the crew could come home without help at the end of the 59-day mission.

Dr. Kraft, director of the space center, got on the radio that afternoon to talk over the situation with the astronauts. Kraft was a cool-headed man who handled many emergencies during an exciting fifteen-year career with NASA.

"We feel fairly confident that two quads will work all right if we have another emergency and have to bring you back quickly," Kraft told the crew in a calm voice. "We have additional small thrusters on the command module as a backup to the service module quads. Should the main engine fail, we feel fairly confident, we could still get you back to Earth. Just to be safe, however, we have started to prepare a rescue mission. We are going to get another spacecraft and rocket ready at Cape Canaveral as fast as possible. Should we need it, we'll have it. In the meantime," Kraft continued, "we'll proceed as if we're going to have a normal mission. Except for giving you some emergency procedures, we would like you to go ahead with your work just as if this were a normal mission."

"You just said the right words," Bean told him happily. "We've been hoping you'd say that ever since we found out there was a true problem with our quads this morning. I think I speak for Jack and Owen, as well as myself, when I say that we're pretty happy with the way things are going at the moment. We agree one hundred percent with everything you've said."

The astronauts went on with their work, while on the ground thousands of people prepared for the first rescue in space.

Rescue Work...
and a
Record
Spacewalk

On August 2, 1973, teams of engineers and technicians at the Kennedy launch center began working twenty-four hours a day, seven days a week, preparing to rescue Bean, Garriott, and Lousma. According to plans made before the emergency, they would use the rocket and spaceship already being assembled for the last Skylab mission, due to blast off in November. The earliest the rescue ship could be ready to go into space was September 5—34 days away.

"There is no danger to the crew that we're aware of, so we are in no panic situation," Bill Schneider said that Thursday afternoon. The second Skylab crew had planned to orbit until September 25, so there was plenty of food, oxygen, and other vital supplies aboard the station. Meanwhile, there was only one thing the astronauts could do. As Chris Kraft had told them: "Go ahead with your work just as if this were a normal mission."

On the ground, feverish activity erupted in many directions. One group tried to determine if leaks in the the two sets of thrusters were connected or separate, and if the other thrusters on the spacecraft might be affected. Another team worked on ways of bringing the astronauts back to Earth without a main engine and only two sets of thrusters working.

A crew to pilot the rescue ship was named—Vance Brand, forty-two at the time, an ex-Marine fighter pilot and civilian test pilot from Longmont, Colorado, and Don Lind, then forty-three, an ex-Navy pilot and space physicist from

SKYLAB RESCUE CSM GENERAL ARRANGEMENT

FOOT
REST

STANDARD COUCH
ASSEMBLY

EXPERIMENT
RETURN
PALLET

AFT BULKHEAD

RESCUE COUCHES

Skylab, the Orbital Workshop, offers long-duration life support in Earth orbit and a practical rescue capability. In the event of an Orbital Workshop failure, a rescue team of two would be launched to return with a crew of five.

Midvale, Utah. Brand and Lind immediately began practicing for the mission in space trainers at the Johnson Space Center.

Carrying Out the Rescue

Here is how the rescue would be made. Men were already fixing up the launch tower after the Bean crew's blast-off. While this work went on, other crews assembled the two stages of the Saturn IB rocket in the huge, fifty-two-story Vertical Assembly Building. With this completed, the spacecraft would be "stacked" atop the rocket. Technicians hoped to have this done by August 10. Next, the 224-foot-tall assembly would be moved to the launch pad on the world's largest tracked vehicle. NASA hoped the move could be made on August 13 or 14. Then would come flight-readiness tests and loading of fuel. The final step would be installation of special rescue equipment.

Storage lockers would be removed from the spacecraft and two extra seats, or couches, put in. These would fit underneath the three regular seats. Brand and Lind would sit in the left- and right-hand seats while flying up to the space station. Two of the astronauts now up there would crawl into the couches below these seats. A third would take his place in the empty center couch. There would be room between the two astronauts "downstairs" to bring back film, tape, and other precious data.

Brand and Lind spent many hours in the simulators at Houston, practicing the maneuvers required to catch up with the space station and fly along close to it. When they actually did this on rescue day, Bean's crew would don their spacesuits and enter the multiple docking adapter. They would close hatches to seal themselves off from the rest of the space station. Once sure that they had a good flow of oxygen into their spacesuits, the three would let the air out of the

MDA. Then Bean, Garriott, and Lousma would use special equipment to undock their crippled spaceship and eject it away from the docking port. Brand and Lind would dock the rescue craft at this port.

If Bean's crew could not get rid of the crippled ship, then the rescue vehicle would dock at a second port on the side of the MDA. But Schneider and other NASA officials wanted to clear the regular docking port. They wanted to use it for the third crew which they still intended to put aboard the space station after Bean's team was safely home. The side port could not be used for this because it had no connections for hooking up the power and communications systems on the spacecraft to those of the space station.

What if the rescue didn't work? Well, there were several other things NASA planners could do. They could try a second rescue using a rocket and spacecraft that they were saving as a rescue craft for the last crew. If the last crew also ran into a problem and had to be rescued, there was yet another vehicle NASA planned to use on a joint flight with the Russians in 1975.

Officials also considered sending up a supply ship if the first rescue attempt failed. This would give them more time to get a second rescue ship ready. There was even the possibility of calling on the Russians for help.

The Situation Improves

By the next day, it was beginning to look as though none of these drastic plans would be necessary. Engineers were coming to the conclusion that the leaks in the two sets of thrusters had nothing to do with each other and that the other thrusters would not be affected. Simulations showed that the astronauts probably could get down all right by using a combination of the big main engine on the back of the service module, the two remaining sets of thrusters on the service module, and the four sets of small thrusters on the command module.

"We're feeling considerably better than we did yesterday morning," Schneider remarked on Friday afternoon. Instead of going full speed toward a rescue mission, the Skylab program director decided to go full speed toward the third mission. If a rescue turned out to be necessary, it could still be accomplished quickly. If rescue was not necessary, then NASA would be ready to send up another crew. The only difference would be that rescue now would be delayed from September 5 to 10.

Ten Thousand Miles in 33 Minutes

Meanwhile, work—and problems—went on as usual aboard the space station. On Friday morning, August 3, the crew did their first survey of the Earth. As Owen Garriott explained it, there is a "window" through which they could look at Earth. In this window were placed six different cameras and other instruments. They looked down at that part of Earth under the space station, taking pictures and making recordings of what the instruments "saw."

A vertical view of Mt. Ranier in the state of Washington photographed from the Skylab space station. Mt. Ranier, in the Cascade Range, is located southeast of Seattle. This is one of many pictures of the Earth taken during the Skylab missions.

On that first survey, the instruments were switched on about 1 P.M., when Skylab was about a hundred miles off the coast of Oregon. The station then swept across Oregon, Utah, Colorado, Mexico, and Texas while Jack Lousma worked to get data on the special "target" sites that scientists wanted surveyed. Skylab crossed the Texas coast and sped over the Gulf of Mexico, across the coast of South America, over Brazil and Argentina and out over the South Atlantic. The station covered almost 10,000 miles in 33 minutes.

"One of the hardest things to get used to up here," remarked Garriott, "is how fast you cover the world. Once around the world in ninety-three minutes and across the United States in the blink of an eye. That's progress—men once took six months to travel from coast to coast by covered wagon or horse."

Just after they went to bed on Friday night, a master alarm woke the astronauts. They rushed around checking the spacecraft and station but could find nothing wrong. Ground controllers checked their instrument panels and found that a short circuit had temporarily knocked out television cameras on the control console for the telescope mount. The television cameras take pictures which allow the astronauts to see what is happening on the Sun and where the telescopes are looking. Fortunately, the station contained a back-up electrical system to take the place of the one that burned out.

On Saturday, August 4, Bean's crew ran a second pass with the Earth Resources Experiment Package (EREP). Sunday, they did two EREP passes, something that hadn't been done before in the Skylab program. You remember that information collected on these passes would be used for such purposes as geological mapping, exploring for oil and metal deposits, determining how land is being used, surveying crops and forests for insect and other damage, studying winds, clouds and storms. For example, two projects dealt with improving storm predictions and warnings in areas where hurricanes and tornadoes occur. Another project involved determining if spacecraft instruments could find and track the movements of game fish such as tuna, marlin, and tarpon.

Long Spacewalk

This Skylab crew had been scheduled to take a spacewalk on July 31, to replace film and cameras in the telescope mount and to put up another sunshade. But sickness and trouble with thrusters kept delaying this extravehicular activity (EVA). Finally Garriott and Lousma received the go-ahead for the space walk on Monday, August 6, their tenth day in orbit. Bean stayed in the spacecraft to keep close check on them and the space station.

It was 12:32 P.M. when Owen Garriott opened the hatch and stepped outside as Skylab streaked over the United States. "It's beautiful out here," he remarked. "What a view!"

"If you'd only get out of my way, I could look," said Lousma.

Jack made his way out to the telescope mount. He fixed himself in a position overlooking the part of the space workshop that was covered by the parasol.

Owen, or "Big O," as they called him, stood in a well-like compartment near the airlock hatch. O's job was to assemble 5-foot-long sections into two 55-foot-long poles. The gradually lengthening poles would be fed to Lousma, about 21 feet away. Jack was to fix the ends of these poles into two plates that he had attached to the workshop "roof." A series of lines and pulleys on the poles would be used to stretch a 24-foot-long, 22-foot-wide shade out to the end of the poles and over the parasol. It worked much like raising a flag or sail except that the shade moved horizontally instead of vertically. Technicians wanted to put up the new sail because they feared that the hurriedly made parasol would not last until the end of the Skylab missions.

It turned out to be slow going. O had trouble getting the pole sections out of

Scientist-astronaut Owen K. Garriott is seen performing extravehicular activity at the Telescope Mount of the Skylab space station. He collected space dust to determine the number and size of particles that hit the stations.

the device which held them. At first, it took him seven minutes to get each rod out and attached to the one before it.

Meanwhile, Jack was hanging upside in the footholds and watching the Earth go by. "The way I'm standing up here," he told the ground, "I'm facing backwards looking down at the Earth. We're halfway down the Lower California peninsula now, and I can see almost up to San Francisco. It's just an all-around beautiful sight."

Controllers thought it would take 2 hours and 15 minutes to put up the sunshade, but after 2 hours had passed only one assembled pole was in place. Then Garriott found a better way to get the poles together and things went faster. After 3 hours they finally got both poles out. Then it was found that the line to pull the shade out was wrapped around one pole. They had to disconnect a section of the pole, then untangle the line so it would slide freely through round fittings on the pole.

But the shade wouldn't unroll or unfurl. It had been carried up to the station by Conrad's crew, and it lay stored in a folded-up position since that time. The folds tended to set and stick together. Bean reported to the Earth that his spacemates were not tired, and they were working steadily on the problem.

After 4 hours outside, Owen and Jack got the sunshade rolled out over the workshop. The parasol had been pulled down close to the roof before the EVA and the shade now covered it. The shade was made out of rip-resistant nylon coated with a special white paint to reflect the Sun's rays. Temperatures inside the station started coming down right away, eventually dropping from the 80s into the 70s.

The rest of the spacewalk went smoothly. Jack worked his way out to a station below the solar panels on the telescope mount. He loaded two cameras with new film passed out by Owen. Next, Lousma made his way to the work station at the top of the telescope mount above the panels and at their hub. Owen sent him two 32-pound cameras and two 23-pound film magazines. Jack put these in place without difficulty.

While out on the end of the telescope mount, Lousma installed four small panels. These contained detectors that recorded the size and speed of small meteorites hitting the station. Scientists and engineers wanted to find out if tiny meteorites that a spacecraft encounters while orbiting close to Earth would penetrate various thicknesses of metal. The detectors were picked up on a later spacewalk and showed that no danger of punctures existed from the usual size meteors.

All the while, Skylab passed from sunlight to darkness and back again on each 93-minute swing around the world. Darkness slowed things down, although the station had powerful outside lights for spacewalking.

Lousma and Owen finally got back inside the space station and closed the hatch at 7:01 P.M. They had been outside 6 hours and 31 minutes—3 hours longer

Closeup view of astronaut Lousma participating in extravehicular activity. While out on the end of the Telescope Mount, he installed four small panels used to detect and record the size and speed of small meteorites hitting the station.

than originally planned and a record for Skylab. They had almost doubled the previous spacewalk record of 3 hours and 23 minutes set by Conrad's crew.

Rescue Called Off

The tremendously successful spacewalk cheered the whole Skylab team. In addition, they were now confident that the leaks in the two quads were not connected and would not affect the other thrusters. Morale aboard the space station improved. No more sickness occurred and the crew caught up with their work. It looked more and more as if they could complete the mission and come home safely on their own.

On the afternoon of August 14, as the next Skylab rocket and spacecraft moved to the launch pad, Bill Schneider and his men reached a decision about the $2-million rescue attempt. "Based on an analysis of the spacecraft which is in orbit and the Skylab workshop," he concluded ". . . there is no imminent need for rescue. Therefore, we have made the decision to use the vehicle that is being prepared for a normal mission. If something else happens, we still can rescue the crew, but we are not planning on it now."

Once again the astronauts and ground teams had worked around a major problem. They had avoided disaster and avoided bringing the mission to an unscheduled end.

The
Jet-powered
Backpack

"With all the things that have gone wrong, are you getting pessimistic about the future of this or the next Skylab mission?" a newsman asked flight director Milton Windler.

Windler replied that teams at the control center were getting optimistic instead of pessimistic.

"I say that because, when we keep having problems and solving them, we get more confidence," he told the reporter. "We're going to try to start catching up. The crew seems to be in pretty good shape, and we're ready to press on with the scientific activities."

Besides studies of themselves, the Sun, and Earth, Skylab astronauts had a number of other tasks, corollary experiments which fit in naturally with what they were doing. These included observation of the stars, attempting to weld and to work with metals in weightlessness and testing a space gun and jet-powered backpack for maneuvering around in space. Also aboard the space station were experiments designed by students in grades 9 through 12. As part of a national contest, announced in October, 1971, 3,409 experiment ideas were sent to NASA and the National Science Teachers Association. Out of these, NASA and NSTA chose twenty-five winners. Six of these experiments could not go into space because they required too much money, equipment or demand on the astronauts' time. Four of the students who suggested these projects received data from Skylab experiments similar to the ones they proposed. The other two had an opportunity to work with NASA scientists in areas close to their major interest. Of the remaining nineteen, eight of the experiments used information which the scientists and astronauts planned to collect anyhow. The final eleven experiments needed special equipment and some of the crew's time.

Spidernauts Spin in Space

One of these involved Arabella and Anita—the two spiders mentioned earlier. Judy Miles of Lexington High School in Lexington, Massachusetts, wanted to determine how a spider might spin a web without any gravity or a sense of up and down to guide it. Some scientists hoped that what was learned about changes in the muscles and nerves of spiders could be applied to humans with muscle and nerve problems.

Common backyard spiders, known as cross spiders, were picked for the trip because they had been well studied on the ground. Experimenters decided that

Miss Judith Miles, Skylab student experimenter, discusses her project with engineers and scientists at the NASA-Marshall Space Flight Center. The experiment equipment involved determining how spiders spin webs when no way exists to tell the difference between up and down.

the spidernauts should be females since they build more webs than males. At the age of four months, well fed with houseflies, Arabella and Anita headed for space in vials carried by Owen Garriott.

He described the experiment results this way: "The first day we opened her vial, Arabella didn't come out at all. The next day, I had to tap and shake the vial to get her into the fifteen-inch-square Plexiglas cage. At first her legs flopped around, and she bounced from wall to wall."

The next day, Arabella clung to the corners of the cage. She spun webs around all four corners and ran threads from corner to corner. On the following day, Arabella spun a more or less normal web.

By the third day of spinning, she had built a good web with the center facing the only opening to her cage. Then Arabella sat back and waited for a pill bug, fly, or other prey.

"She learned very rapidly," Garriott commented. "Without the benefit of previous experience, she figured out a nice solution to the problems of zero gravity. After her two- to three-day adaptation period, she seemed to enjoy zero-g just as much as the three of us did."

Arabella spun away until August 27, when she was replaced by Anita. After a short period of confusion, Anita got her webs straight and seemed to have no difficulty living aboard Skylab. However, on September 16, Garriott found her dead in her cage. The scientist-astronaut figured that Anita did not eat the fly-size bits of filet mignon the astronauts contributed from their food and she died of starvation. Arabella thrived on the steak diet and came home with the astronauts. But when her vial was opened at the Marshall Spaceflight Center, she, too, was dead.

Something Fishy

The two minnows Garriott brought with him learned to swim normally without knowing which way was up or down. At first, they swam in confused circles in a plastic bag filled with sea water. Instead of swimming horizontally, they made loops with their heads tilted up or down. Garriott hung the bag against a locker door in the living quarters. Gradually, the fish came to accept the dark, solid locker surface as the down direction—the "bottom" of their "pond." After this they swam normally, but still became easily disoriented when someone shook their bag.

Garriott also carried fifty minnow eggs up to Skylab and forty-eight of them hatched. The space-born babies were not confused by the lack of gravity at all. They immediately adopted the locker wall as "down," and swam normally from the day they were born. No amount of shaking their sea-water bag caused them to lose their sense of position.

According to space veterinarian Dr. R. C. Simmonds of the Johnson Space Center, the older fish used their eyes and brains to replace those senses of

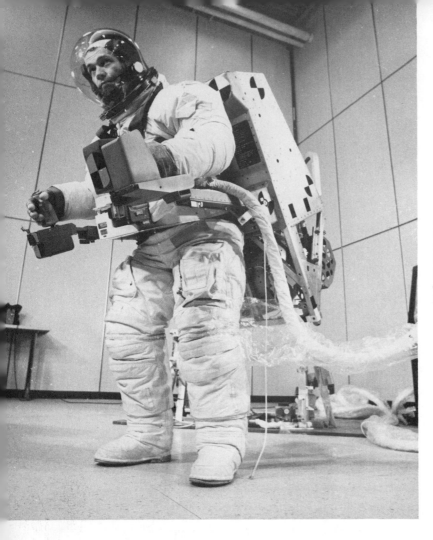

Astronaut Bruce McCandless tests the backpack for use in the orbiting space stations. Powered by nitrogen gas jets, the unit has four different control systems to evaluate maneuvering in weightlessness.

balance and direction which do not operate in zero gravity. In a way, this is what Arabella and the astronauts did. After a period of adjustment, sometimes marked by sickness, a man's eyes and brain take over. He accepts one part of the spacecraft as up, another as down, then goes about doing his work. The same would probably be true for women. But would human babies born in space adapt immediately to floating around weightlessly, as baby fish did? The answer to that will have to wait until the first baby is born in space.

Jetting Around the Spacehouse

One of the most interesting jobs for the second Skylab crew was piloting various one-man flying machines. Powered by small jets, these devices were designed to test various ways future astronauts could fly from one spacecraft to another, or do outside work such as spaceship repair or building a large space station.

One machine, called an astronaut maneuvering unit, consisted of a seat with arms mounted on a large backpack—41 inches high, 27 inches wide. The pack held a bottle of compressed nitrogen gas, a battery, and 7 pairs of thrusters. Each arm held a hand-grip control of the same type as the Apollo spacecraft. By moving these controls, the astronaut operated the thrusters and flew around in the same way as they would in the Apollo ship. This unit was, in fact, a one-man spacecraft.

On August 13, Al Bean tried flying this spacecraft around the upper floor of the workshop, which was about 20 feet high and 22 feet across. Wide open, the thrusters could build up a speed of 27 mph. But Bean was instructed to keep the speed down to .3 mph, slower than you would walk. This speed enabled him to fly around the cabin without hard bumps into protruding equipment and lockers.

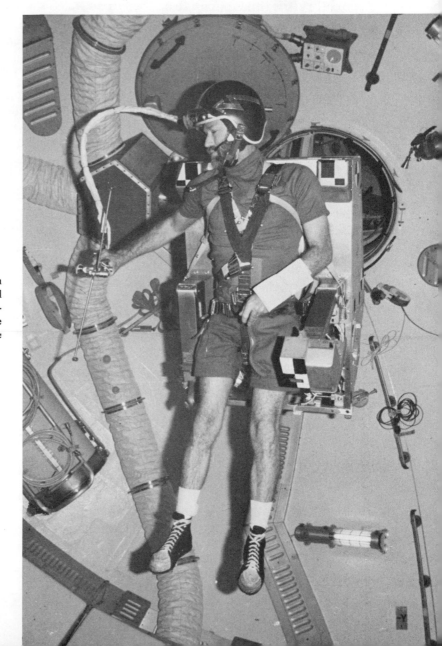

Astronaut flies the one-man spaceship using hand-held push-button controls. The experiment took place inside the forward compartment of the OWS.

"It's very, very easy to fly," Bean reported. "Flying it around the workshop is quite simple; there's no problem."

Lousma also tried flying with the backpack. "It flies very well," he told the ground, "like most any spacecraft."

Bean had one complaint. When maneuvering outside, astronauts are attached to the ship by an umbilical—a thick tube containing lines for cooling water, oxygen, power, communications, and instrumentation. Inside the station, this stiff umbilical twisted and shoved Bean in directions he did not want to go. He had to use a lot of fuel to "overpower" the umbilical.

Trying the Space Gun

On August 15, the commander flew the one-man spaceship again. He also tried flying around with a hand-held maneuvering gun. This consisted of a hand grip with button controls and two 15-inch barrels sticking out on either side. Its "ammunition" was nitrogen gas supplied by a short hose connected to the backpack. To use it, Bean folded the arms and controls of the astronaut maneuvering unit out of the way. He held the gun in his right hand, faced where he wanted to go, aligned the "space gun" with the center of his body, and pressed one of the triggers. To stop or go backward, Al pressed another button trigger.

Such maneuvering guns had been used outside spacecraft on missions in 1965 and 1966, but they did not work out too well. Bean also complained about this space gun. If he changed his body position while the gun was firing, it threw him off course and he found it hard to get headed in the right direction again.

"I could fly it and didn't bump into anything," Al said. "But it's like riding a unicycle. You can do it, but it isn't the sort of thing you want to do if you need to get somewhere, especially if you are working outside a spacecraft. It just doesn't make good sense to go outside with a thing that requires skills that you don't normally use."

A couple of days later, Bean and Lousma tried flying the one-man spaceship with their spacesuits on. Again, the umbilical gave them a problem. "Without the umbilical, it flies very well," Bean reported. He recommended that the best way to fly the craft outside would be to replace the heavy umbilical with a light, flexible safety line, combined with light oxygen and communication lines. He wanted a life line that was not so stiff it pushed, pulled, and twisted an astronaut.

Flying with Jet Shoes

Bean tried "jet shoes" next. With the backpack and its tank of gas strapped on, Bean sat on a saddlelike seat. At the base of this were "stirrups" holding foot pedals and two quads of tiny thrusters. By moving his feet and toes, Bean could fire different combinations of thrusters and fly around the space station. The biggest advantage of jet shoes was that they left the astronauts' hands free to do work or to carry loads. The biggest disadvantage was that you could move headfirst or feetfirst in any direction, but you could not go from side to side, or back and forth if you stayed upright.

When you're strapped into this foot-controlled maneuvering unit in a spacesuit, "you're in a semihelpless situation," Bean reported. "You have to be pointed right at the target. Sometimes you can make it, but once you get there, you can't move your body too well. Your arms are free, but it's like sitting on the back of a convertible seat and trying to drive the car with your feet. You can't get to the brakes fast, and you can't get to the clutch. If you have any failures, you're out of business. Maybe it's good to test it, but I don't think you'd ever want to fly it for real."

After the mission, Bean summed up the crew's experience with one-man spaceships, space guns and jet shoes this way: "The maneuvering unit is quite reliable and easy to fly. After I flew it and Jack flew it, we wanted Owen to try it. With no previous training on the ground, such as we had, he flew it just as well as either of us. I believe that this sort of maneuvering unit, built lighter and smaller, could be used with minimal training, and it would do the job you wanted to do outside a spacecraft."

The Skylab commander added that he would not want to work outside a spacecraft with either the foot-controlled or hand-gun unit.

What uses would these one-man spaceships have in the future? I put this question to Air Force Major C. E. "Ed" Whitsett, the man in charge of the astronaut maneuvering unit and space gun projects.

"We see a large number of possibilities in the 1980s and beyond," he answered. "By then, space shuttles should be carrying men, satellites and equipment into space, and bringing such loads back. Astronauts might need to transfer themselves and their equipment between spacecraft. They might fly over to unmanned satellites to do repair or maintenance work, such as changing batteries or experiments. They might want to fly around the outside of a space station to do repair work. We sure could have used this capability on the Skylab missions. Maneuvering units could be vital for rescue work, or for getting a disabled satellite aboard the shuttle so it could be brought back to Earth.

"In the near future, when astronauts work fairly close to the mothership, we'll probably use a light, flexible umbilical. Later, maneuvering units would have their own oxygen, power, and communication supply, though they still might have a safety line. When we fly out to distances of a thousand feet, astronauts probably will do away with lines connecting them to the mothership. Just as astronauts walked and drove around the Moon, someday they will move about in space on their own."

Chapter

An
Exploding
Sun

On his eighth day aboard Skylab, Al Bean complained that the astronauts didn't have enough time to finish their work each day.

"There hasn't been a night that we didn't get to bed after eleven o'clock," he radioed the ground. "Although you guys have been asking about the sights out the window, I'll bet our total looking-out-the-window time has been less than two hours for all three of us together."

However, by the sixteenth day, when the motion sickness was over and problems aboard the space station were under control, the situation changed.

"Now we end up getting finished early in the day," Bean said. "It's taking us only thirty minutes or less to eat." (The flight plan allotted one hour for meals.)

The Skylab commander asked for more work. He decided to start an hour earlier in the morning and stop an hour later at night. This meant a fifteen- to sixteen-hour workday for the astronauts.

"None of us around here wants to waste any of this valuable time because everybody has got too much invested in the mission," Bean said. Lousma and Garriott agreed.

The crew put part of the blame on themselves for getting off to such a slow start. The astronauts had requested that they start off doing the same amount of daily work that Conrad's crew did in the middle of their 28-day mission. Bean's team thought this would be possible because they had learned how to do things and how to avoid mistakes from the first crew.

"But you can't pull that off on day two, three, or four because you lose things, and you don't even know how to stand in one place," Bean observed. "My guess is that the first time you do anything, you ought to have more time." (Mission planners followed this advice by allowing the third crew 50 percent more time to do experiments the first time.)

When they began working at a faster pace, Bean, Garriott, and Lousma did more jobs than had been planned. And they had fun doing them. "I personally have had more fun on Skylab than on the lunar mission [when he landed on the Moon]," Al Bean said. "Flying weightlessly around this workshop is really a ball. In the command module on a trip to the Moon there's nowhere to go. Here you can zoom forty feet, and you don't even have to go by handrail. You just push off in the direction you want to go."

Another "Fix-It" Spacewalk

Despite the high morale and high level of activity, problems were not over. Since the first manned mission, controllers had been worrying about the gyro system that sensed movements of the station and kept it flying straight and level. This consisted of nine gyroscopes. Three sensed when the station tilted up and down, three when it rolled, and three sensed yawing from side to side. Only one gyro was necessary in each case, but engineers put three aboard for extra protection.

It was a good thing they did. Part of the gyros failed on Conrad's mission, and others acted erratically during the present mission. By the fourth week, only one gyro in each package of three worked right. And one of these was starting to act up. Bean's crew had brought a "six-pack" of new gyros with them because of problems on the first mission. Flight planners decided that the astronauts should install these gyros when they went outside to change film and cameras on August 24.

The gyros work just like part of your own nervous system. When you trip or start to lose your balance, your nerves send warning signals to your brain. Your brain, in turn, sends instructions to other nerves and muscles that keep you from falling. The gyros sensed unwanted tilting, drifting, and other movements, and sent signals to a computer in the telescope mount. The computer calculated the force needed to stop this motion and sent signals to large spinning gyroscope wheels in the ATM. Movements of these wheels and the firing of thrusters on the outside of the workshop restored the station to straight and level flight.

The small sensing gyros that needed to be replaced were mounted inside the MDA. But the astronauts had to go outside to replace a cable carrying gyro signals from the MDA to the computer. Jack Lousma, a powerfully built, six-foot Marine, got the job. He stepped outside the hatch at 11:24 A.M. while Skylab orbited over the South Atlantic.

"Oh, boy, there's the world," he said, looking down.

Spacesuited astronauts practice loading and unloading cameras on mock-up of Skylab Telescope Mount. Mount is placed underwater to simulate weightless conditions of space. "Dutch shoe" restraints hold astronaut in place while he is working.

Lousma unplugged the old cable with a pair of pliers. The new cable was 23 feet long. He connected one end to the outside of the MDA and the other to the computer on the telescope mount. While he did this, Al Bean put in the new six-pack. The task went smoothly. At 1:30 P.M. the flight director reported the new gyros were "doing great."

When Jack finished, Owen moved out to the near end of the telescope mount. He changed film sent back and forth on the extendable boom by Lousma. Next, he moved out to the far end of the telescope mount and replaced film and cameras there.

Doors which protected the lenses of two instruments were not working right. One was sticking closed at times and the other did not close properly. The doors were not really needed out in space to protect the lenses. Therefore, Owen removed both of them.

Garriott and Lousma got back inside their spacehouse 4 hours and 31 minutes after starting their spacewalk. They had lived up to the "we fix anything" motto of Skylab.

"The EVA has got to be termed an unqualified success," flight director Neil Hutchinson said happily.

Breaking Another Record

The day after the spacewalk, on August 25, at 7 A.M., Bean's crew broke the first Skylab crew's record of 28 days, 49 minutes in space. Then on September 5, Bean broke Pete Conrad's record of 1,180 hours to become the world champion of space. Conrad had accumulated 49 days, 3 hours, 37 minutes in space on two Gemini flights, a Moon landing, and the first Skylab mission. Bean had flown on the nine-day Moon trip with Conrad and he had been aboard Skylab more than forty days.

"I guess I'll be handing that record over to the next Skylab crew in three or four months," Bean told the ground controllers when they congratulated him.

In addition to the excitement of breaking records, astronauts get a big thrill when they fly over their hometowns. This happened to Jack Lousma the night of September 6. As Skylab passed over Michigan, Jack radioed: "I want to say hello to all of the folks who live in Grand Rapids. I've got lots of relatives and friends down there. Grand Rapids is the place where I was first launched [born], and I've got lots of good fun memories from the time that I lived there. As we whistled over the city, it was a clear night. I want to thank all of the folks for flashing their lights—just don't send me your electric bill."

The Sun Erupts

On the morning of August 21, it was a fierce and explosive Sun—not the friendly people in Grand Rapids—that flashed its light at Skylab. Astronomers watching the Sun at an observatory in the Canary Islands and monitoring instruments on an unmanned satellite, radioed Skylab control about disturbances on the eastern edge of the star. Controllers quickly relayed word to the astronauts. Orbiting above most of Earth's hazy atmosphere with the large powerful instruments, they would get the best view of what was going on by means of television pictures and filtered images.

"You can see a big bubble sitting on the edge of the Sun," Bean reported at about 9:45 A.M. "It's about three-quarters the size of the Sun and it's expanding outward."

This great bubble of hot, electrified gas broke free of the Sun and hurled through space at a speed close to a million miles an hour. The mass of gas was extremely thin but hundreds of times bigger than the entire Earth.

The electrically charged gas traveled across the 93 million miles between the Sun and Earth and slammed violently into our planet's upper atmosphere.

Such an encounter is as powerful as a hurricane. But, fortunately, Earth is surrounded by a magnetic field which stops and absorbs the gas particles and radiation. We neither see nor feel, directly, the violence that takes place hundreds of miles above us.

However, this does not mean that these magnetic storms have no effect on us. The charged particles from the Sun interact with those in our own atmosphere and produce spectacular displays of shimmering color, called northern

Charles Conrad, Jr., commander of the first manned Skylab mission, is shown at the Apollo Telescope Mount console. Astronomy experiments increased knowledge of the Sun and its effects on Earth.

and southern lights. We depend on our magnetic fields to transmit radio waves over long distances. When solar activity disturbs these fields, long-distance radio communications sometimes are blocked out. Unwanted electric currents build up in power lines and cause blackouts. Some scientists believe such events also effect our weather.

The Sun frequently ejects vast quantities of energy and material. Skylab provided scientists with the best photographs and information on solar eruptions to date. The one on August 21, for instance, could not be seen clearly from Earth, even with the most powerful telescopes.

Solar eruptions occur in the corona, or outer atmosphere, starting about 6,000 miles above the solar surface. This part of the Sun is so thin and weakly lighted that the glare from the surface makes it impossible to see. For many years, scientists could study the corona only when an eclipse blocked out the main part of the glaring disc. You may have seen photographs of a Moon or Earth shadow covering the Sun. The halo and rays of white light that appear around the edges of the eclipsed Sun are part of the corona. Aboard Skylab, an instrument called a coronagraph constantly blacked out the main body of the Sun, creating a continuous artificial eclipse. With this instrument a complete photographic record could be made of what went on in the corona.

Of course, you can use a coronagraph on the ground, too. But in this case, you are at the bottom of an "ocean" of cloudy, dirty, moving air, hundreds of miles thick. Not only does this obstruct visible light but it absorbs other types of radiation, such as ultraviolet and x-rays. These rays cannot be seen by the naked eye, but they reveal a great deal about what goes on in the Sun. Scientists want to understand as much as possible about this because all life on Earth depends on the light, heat, and other energy that our star provides.

Although it makes things tough for astronomers, it is fortunate that all solar x-rays and most ultraviolet radiation is stopped by our atmosphere. Many scientists believe that too much exposure to x-rays produces cancer, while unprotected exposure to ultraviolet rays results in deadly burns. The astronauts worked above most of the atmosphere, but Skylab's metal gave them enough protection. On spacewalks their spacesuits protected them. The astronauts could safely record information never obtained before with the eight different instruments on the telescope mount. In addition to the coronagraph, two recorded x-ray activity and three recorded ultraviolet radiation.

Another instrument consisted of two telescopes through which television pictures and photographs were taken of the red light given off by hydrogen. All other types of light were eliminated. The Sun consists mainly of hydrogen, and this instrument produced detailed photos of solar activity invisible to the eye and ordinary telescopes.

Never before had astronomers obtained such detailed and continuous views of the Sun. Before Skylab, Sun-watching was like trying to see a football game from a high-flying airplane. You could spot the field through the clouds when the plane flew over, but you couldn't see the players or follow the plays. With Skylab, scientists obtained a detailed view of the "players" and events on the Sun that mean so much to us on Earth. Information coming out of these solar observations has been compared with the new knowledge gained when man first looked at the heavens through a telescope. Scientists said that the new discoveries produced by this information were as exciting as those about the Moon that resulted when men first landed there.

Flares Release Tremendous Energy

The eruption on August 21 was followed by a period of intense flare-ups on the Sun. All had been quiet on the star during much of the year. Then, suddenly, bright, active areas, called flares, began appearing. These sudden bursts of light signaled the release of tremendous amounts of energy. During the afternoon of September 5, two flares lasting from ten to twenty minutes sent the astronauts rushing to the controls of the Apollo Telescope Mount. An Earth survey was canceled so the crew could keep close watch on the Sun. A huge tongue of hot gas surged out of one area and looped high into the corona. It traveled some 100,000 miles and came down in another area, which then began to flare.

That same day Jack Lousma spotted aurora, or southern lights, flashing in the sky above Australia. Later, Al Bean told mission control that they saw aurora at both ends of the Earth in the same day. "We saw the northern lights and the southern lights all in one day, which may be a first for anybody," he said excitedly.

The next morning, while the astronauts slept, another major outburst occurred on the Sun. A flare on the western edge, together with other radiation, hurled out ten times the amount of x-rays yet seen on Skylab. Commands were sent up to the ATM and the instruments automatically began taking photographs and readings as often as every fifteen seconds. Al Bean got to the television monitors about 6:30 A.M. An hour later he reported that a bright streamer from the flare shot out hundreds of thousands of miles into black space.

"A typical flare like this produces about a hundred million times more energy than the biggest atom bomb or earthquake," noted Joseph Hirman, a solar scientist with the National Oceanic and Atmospheric Administration. He added that anybody traveling in space beyond the protection of Earth's atmosphere, would have been made "pretty sick, pretty fast" by the radiation. If Alan Bean had been exploring the Moon again instead of flying in Skylab, he would have been forced into the lunar landing craft for protection. Even then, he might not have been safe. In such a case, Al would have to blast off from the Moon and rendezvous with the Apollo spacecraft that orbits overhead. That ship has more protective shielding.

Next day, September 7, an even bigger flare broke out. It was the largest and brightest one that had occurred to date on Skylab. It was ten times the size of the Earth, and it sent charged particles zooming through space at speeds of 35,000 miles a *second*. Thirty minutes after the flare started, at about 7 A.M., northern lights flashed on over Earth and long-distance radio communication completely faded out for twenty minutes.

The outburst got scientists very excited. First reports of it came about thirty minutes before the astronauts were supposed to man the ATM controls. Skylab was over Guam in the Pacific, and scientists could not get word to them before the station passed out of radio range. The next contact was from the USNS *Vanguard* in the South Atlantic, but because of a problem no voice exchange could take place.

"We were faced with the situation where we had no way of giving the crew any instructions," said James Milligan, a NASA physicist. "There was a lot of hair pulling on the ground hoping the crew got to the control panel and did the right thing."

They need not have worried. Jack Lousma began training the telescopes on the Sun the minute it rose that morning.

"It's the big daddy," shouted Bean when he saw the flare.

"When you look at the Sun now, it looks like someone kicked the heck out of it," exclaimed Garriott.

All the scientists were extremely pleased with the performance of the astronauts. "I think it really demonstrated the advantage of having a man in space and being able to react in time," said Milligan. "The crew did a beautiful job and observed the flare just exactly the way everybody wanted them to do it."

And what a flare it was! Hirman calculated that "you could probably run everything in the U.S. and the rest of the world combined for five hundred years on the energy that was given out in this individual flare."

No one knows what produces these gigantic outpourings of energy from our mother star. They occur in association with the dark sunspots that appear and disappear about every eleven years. For years, scientists will study the thousands of photographs and miles of data brought back by the Skylab astronauts to learn more about flares. Understanding what causes them is the first step toward predicting when they will occur. This would be invaluable for forecasting radio and power blackouts, times of danger for high-flying airplanes and even long-range prediction of weather.

Scientists will search Skylab data for answers to other mysteries about the Sun. Most of these have to do with the star's source of energy. It is a huge nuclear furnace, 332,000 times as massive as Earth. Intense heat, as much as 27,000,000° F. at the center, fuses hydrogen atoms into helium, releasing tremendous energy in the process. This is the same nuclear reaction that gives a hydrogen bomb its power. But the bomb's force is explosive, while the Sun's burning of itself is controlled.

Magnetic fields play the major role in making the Sun a furnace rather than a bomb. These fields hold intensely hot, electrified gas in place. When the pressure of this gas exceeds the holding force of the magnetic field, vast blobs of material break away from the Sun, as happened on August 21. Flares are thought to be caused by the breaking, or twisting, of the magnetic field. One solar scientist compares it to the twisting of a rubber band until it suddenly snaps, producing a rapid release of energy.

Scientists want to know how to confine such a hot, electrified gas with a magnetic field. They have decided to try to reproduce in the laboratory the conditions that Skylab astronauts observed on the Sun. If they can learn to do this, they can make a hydrogen bomb burn like a furnace—one that produces vast energy for peaceful purposes. The idea is to learn enough about the Sun to make a mini-star on Earth. A constellation of such stars would give man unlimited energy with which he could make this a better world for everyone.

"Going Back to Houston"

At two o'clock in the morning, as Bean, Garriott, and Lousma orbited over Australia, they heard Dean Martin singing "Going Back to Houston." This was the ground controllers way of waking them up on their last day in space.

"Watch out, Houston, here we come," exclaimed Al Bean as he heard the record on the Skylab radio.

The early wakeup on September 25 marked the beginning of a long, eventful day for the astronauts. Their job was to bring themselves and their crippled spacecraft down to Earth. Two problems faced them:

First, with two of four clusters of rocket engines not working, the two remaining clusters needed to be test fired. If they didn't work, NASA officials might have to send up a rescue team. The rescue rocket still stood on the launch pad. It would take about a week to fuel it and get ready.

Second, was their physical condition. After 59 days in space, their bodies had adapted to the lack of gravity. Doctors feared that their hearts and muscles had become weaker and that their hearts could not pump enough blood to their brains against the pull of gravity, causing fainting, or worse.

On top of these worries, Bean, Garriott, and Lousma had to eat a cold breakfast. The astronauts turned off power to the food warmers the night before as they closed up the station. Not only were the eggs cold but their corn flakes were soggy. The cereal had to be prepared the night before, prior to turning off the water.

After breakfast, Bean crawled into the Apollo spaceship to turn on and check all the systems. Owen and Jack busied themselves carrying blood samples, feces, urine, film, and tapes into the Apollo ship. Three days before, during a 2-hour, 45-minute spacewalk, Al and Owen had unloaded and loaded the telescope cameras again. They would be operated from the ground until another team of astronauts manned the station.

With their ship loaded, Bean prepared to "hot fire," or test, the two good clusters of steering rockets. "We're ready when you are," the controllers told him. "Let her rip!"

Bean fired the rockets. He and the ground teams nervously watched instruments for the results. Seconds ticked away. Then, at about 10:23 A.M., word came from the Houston control center: "Your hot fire is GO. Everything looks good to us."

That problem was solved. The astronauts closed the door of the darkened spacehouse and got ready to come home.

A Dizzy Homecoming

Inside the cramped spacecraft, the astronauts removed their heavy spacesuits. Over cotton underwear they donned "counterpressure garments" designed to protect them against Earth's gravity. The garments reached from ankles to waist. The astronauts used a rubber bulb to pump air into them until the pants felt tight around their legs. This pressure counteracted gravity pulling blood down into their legs. Their blood vessels needed help to prevent blood from collecting in the legs. The upward push of the inflated pants made it easier for the heart to pick up enough blood to supply needs of other parts of the body.

They also took seasick pills, remembering how the motion of the spacecraft bobbing on the ocean had affected Joe Kerwin at the end of his mission. They wanted no part of that feeling.

At 2:48 in the afternoon, as Skylab streaked over the Indian Ocean, Bean separated the Apollo craft from the space station.

"It seems like we're leaving home," said Lousma.

Despite the thrusters that didn't work, Bean reported the spacecraft "flies awfully good—better than I expected."

The astronauts flew once around the world again. Then, just before sunrise over Malayasia, they turned their ship around and fired the big main engine on the rear. An 18-second burst slowed the spacecraft and sent it on a long curving path toward Earth's thick atmosphere.

At 5:04 P.M. they began their entry into Earth's air over the North Pacific Ocean, about 2,000 miles west of Portland, Oregon. During the rapid plunge through the dense air, the pull of gravity increased to 3.8 times its normal force. Bean's weight of 147 pounds increased to 558 pounds. Lousma, weighing close to 190 pounds, was pushed back into his seat with a force of about 720 pounds.

Heat generated by friction with the air made the ship's metal glow and started burning the heat shield away.

"Man, what an entry!" Lousma said to Garriott later. "You should have seen that fireball."

The parachutes opened, and the spacecraft hit the surface of the Pacific at 5:20 P.M.

The astronauts landed 230 miles southwest of San Diego and only 6 miles from the recovery ship, the aircraft carrier USS *New Orleans*. Working in winds of almost 15 mph and waves 6-feet-high, the recovery team of swimmers, helicopter pilots, and ship's crew took only 43 minutes to get the spacecraft onto the deck of the carrier.

The astronauts hoped that they would be strong enough to walk from their ship to the laboratory where they would undergo a six-hour medical examination. However, doctors did not want them to do this because they feared it would spoil some of their medical data. As it turned out, the astronauts felt unsteady and had to be helped the few steps from the spacecraft to waiting chairs. The chairs were on a platform moved by a forklift truck. When they stood up to go into the laboratory, Lousma stumbled against a railing.

All three spacemen were pale and dizzy, but doctors said they were in good physical condition. "There is no evidence of motion sickness in any of the crew," said a medical summary. "Blood pressure and heart rates are within normal limits. There is some evidence of vestibular disturbance [loss of balance and dizziness]. They are in excellent spirits and they feel well."

The Best to Date

Bean, Garriott, and Lousma established a new record for time in space—59 days, 11 hours, and 9 minutes. They traveled over 24 million miles and circled Earth 859 times. The flight was not only the longest but the most successful to date in terms of scientific work accomplished.

"We met the objectives of the mission by well over a hundred and fifty percent," Skylab director Bill Schneider said happily.

The mission, according to Dr. James C. Fletcher, head of NASA, was "one of the most significant scientific ventures of all times. In the fifteen years since it was founded, on October 1, 1968, NASA has successfully placed two hundred fifty payloads [manned and unmanned missions] into space. This mission will very likely prove to be the most fruitful of all these two hundred fifty. The crew completed half again as many Earth resources passes and solar observations as had been planned."

President Nixon sent Bean, Garriott, and Lousma a message of congratulation: "At the completion of mankind's longest journey beyond the boundaries of human knowledge, I congratulate you on behalf of all of the American people. The record of your Skylab mission combines the traditions of those great explorers of history who have faced the uncharted reaches of the physical unknown with the traditions of those men of science who have unlocked the secrets of the universe."

It was a magnificent triumph. But what kind of changes had 59½ days of weightlessness produced in the astronauts' bodies? How long would it take them to get back to normal? Another Skylab crew anxiously waited to go aloft in November. Would doctors find harmful changes that would limit this flight and all flights afterward?

Getting Back to Normal

The astronauts' first physical exam included runs on the lower body negative pressure device, the same instrument on which they tested their hearts almost everyday in space. As described earlier, this consists of a horizontal cylinder into which a spaceman slips up to his waist. Pressure is dropped inside the cylinder, and the heart must work harder to keep blood from pooling in the legs. When pooling prevents blood from getting back to the heart, the heart begins to beat faster. If this increased pumping action doesn't get enough blood up to the brain, fainting occurs. Doctors agreed that the inflated pants helped protect the astronauts against this. Bean could not complete his first heart test on Earth without the garment.

The three also pedaled a wheelless bike, as they did in space. This measured how much work they could do. Bean and Garriott went through these tests all right. Lousma, however, began to feel sick and could not finish his bike run. After the exam, all three could walk without help, but they felt weaker than before the flight.

Next day, the three underwent another seven-hour medical check. Although their hearts beat faster than normal during the bike test, they put out more work than on the previous day. This meant the astronauts already were starting to adapt to Earth's gravity.

On September 27, Dr. W. Royce Hawkins reported that Bean's crew was in better condition after their flight than Conrad's crew. "Their overall condition and general health slightly exceeded our expectations," he said. "We expected that they would look more like Joe Kerwin."

After five days back on Earth, the astronauts' hearts and muscles were close to normal again. It took Conrad's crew twenty-one days to reach this point. Doctors felt that the increased amount of exercise—from a half hour to an hour—contributed a lot to this fast recovery.

After the fortieth day, almost no changes occurred in their muscles or the rest of their bodies, according to Dr. Hawkins. "Somewhere around the thirty-ninth or fortieth day, the astronauts reached the point that there was hardly any change that you could detect," he said.

The chemical and mineral balance of their bodies became stable, too. One

A Saturn 1B launch vehicle is rolled to Launch Complex 39, Pad B. The space vehicle launched by the Saturn 1B carried the third Skylab crew to the orbiting space station.

of the big worries—the loss of calcium from bones—turned out to be a false alarm. "The problem was overrated," admitted NASA doctor Michael Whittle. "The loss we saw in Skylab was not enough to weaken their bones."

The astronauts themselves felt little change in their condition after twenty-five days in space. "After twenty-five days we seemed to hit a groove," said Bean.

"After that point, we thought that we could have stayed up there indefinitely," Lousma added.

Bean commented: "If the next crew stays twice as long in space, I believe that they will recover as fast or faster than we did. The things that established our happiness and well-being up there were getting exercise every day, eating meals on a regular schedule and sleeping between six and seven hours."

The others agreed and recommended that the next crew increase their daily exercise from one hour to one and a half hours.

The doctors, looking at the second crew in more detail, concluded that changes in their bodies had leveled off after about forty days in space. The changes involved in this adaption to living in weightlessness caused no major problems when they came back to Earth. After five days, Bean's crew was much closer to their preflight condition than Conrad's team. Bean, Garriott, and Lousma still felt weak and got dizzy if they moved their heads too fast, but their conditions improved daily.

A New Crew Gets Ready

Elated over the good condition and fast recovery of the second crew, NASA considered sending the third crew into space for as long as 84 days. Gerald Paul Carr, the forty-one-year-old Marine chosen to command the mission, was sure that this would happen.

"One look at Jack Lousma as I tried to follow him running around the track indicates that extending our flight will be no problem," Carr told me on October 2. "He ran a mile and a quarter faster than I did."

The space agency announced on October 26 that it would attempt to launch the last Skylab mission on November 10. The flight would carry enough food, water and other supplies for 84 days. Flying with Carr as pilot would be Air Force Lieutenant Colonel William Reid Pogue, forty-three, from Okemah, Oklahoma. Bill Pogue had flown 43 combat missions in the Korean War and had been a member of a famous Air Force acrobatic team, the *Thunderbirds*. Scientist-astronaut Edward G. Gibson, thirty-seven, an engineer-physicist from Buffalo, New York, was the third crew member. It would be the first trip into space for all three.

They would fly the rocket that NASA had hurriedly prepared for a possible rescue mission. The "bird" had been sitting on the launch pad since August 14. On October 23, technicians pumped 43,000 gallons of a kerosene-type fuel into four of its tanks during a test. This kerosene, together with liquid oxygen, enters the combustion chambers of eight rocket engines. The fuels ignite on contact, giving the first stage of the Saturn IB its explosive thrust.

After the test, as workman drained the fuel, it looked like Skylab's record for trouble was holding: The tops of two of the tanks buckled inward. It had started to rain, and workmen put plastic covers over inlets through which outside air flows into the tanks as they empty. Without this air, pressure in the tanks remained much lower than that outside. The difference was enough to cause the domed-shaped tank tops to buckle inward.

Engineers feared they would have to roll the rocket back inside the assembly building and replace the tanks. Before doing this, however, they decided to try another plan. Technicians pumped kerosene back in the tanks under pres-

Third Skylab crew paused in front of their Saturn 1B rocket at Launch Complex 39B. They are (from left to right) Dr. Edward G. Gibson, scientist-astronaut; Gerald P. Carr, commander; and William R. Pogue, pilot.

sure. They hoped that this would push the pliable aluminum alloy tanks back into shape. The plan worked. The Skylab tradition of "we-fix anything" still held.

Cracks Cause Delays

Astronauts Carr, Gibson, and Pogue passed their last major physical exam on November 6, and they looked forward to a launch four days later. But things did not work out that way. On November 17, technicians discovered tiny cracks in the eight fins of the 20-story-tall rocket. The cracks occurred around bolts which attach the fins to the rocket body. A total of fourteen cracks ranged in size from less than an inch to one and a half inches. The fins support the 1.3-million-pound weight of the rocket as it sits on the pad. After launch, they prevent it from twisting or tumbling in flight.

Engineers believed that exposure to the salt air for almost four months

Technicians remove the first of eight damaged stabilization fins from the Skylab rocket. Cracks discovered during a routine inspection prompted rescheduling of the third mission.

produced some corrosion. When 279,000 pounds of kerosene were pumped into the rocket, the weight caused the corrosion-weakened fins to crack. The additional stress of knifing through the air after launch could cause a rocket cracked in this way to break up, said Walter G. Kapryan, director of launch operations.

There was only one thing to do—delay the launch and put on new fins. Trucks brought the fins from the rocket assembly plant near New Orleans. However, a task like this had never been tried before on the launch pad. Each fin weighed 474 pounds, was 12 feet high and extended 9 feet out from the rocket. Raising the heavy, bulky fins 150 feet and putting them in place was a tricky and dangerous job. Despite gusts of wind up to 33 mph and chilling temperatures, however, the job went ahead. NASA looked forward to a November 15 launch.

As workmen struggled with the fins on November 12, engineer Royce English went about a routine inspection of the rocket. He checked beams in the circular band that connects the first and second stages of the rocket, using a

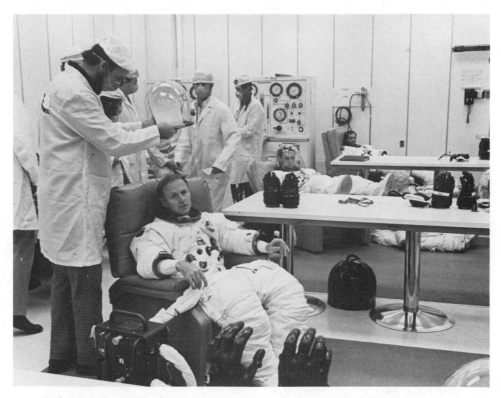

Technician Dan McCarthy prepares to place helmet on head of Commander Carr during spacesuit fit checks a few days prior to his scheduled launch to the orbiting space station.

magnifying glass. What he saw made his heart sink. Cracks as much as two and a half inches long snaked along seven of eight beams.

Engineers quickly inspected a spare rocket in the assembly building at the Cape Kennedy launch center. This was the rescue vehicle. It, too, had cracks in the interstage beams. Richard G. Smith, program manager for the Saturn, gloomily estimated that the first stage of the rocket contained about 2,000 parts which also might be cracked. Most of these parts would have to be checked before the astronauts could ride the rocket into space.

On Tuesday, November 13, just two days before the expected launch, it was called off again. "We want to check over other areas of the rocket for possible cracks," Bill Schneider explained. If more were found, the rocket would have to be moved back into the assembly building for major repairs.

Chapter 11

Sickness, Spacewalking, and Student Experiments

Workmen finished replacing the eight fins on the Skylab vehicle at 6 A.M. on November 13. By that evening, most of the rocket parts that might have suffered stress damage or corrosion had been inspected. No additional cracks turned up.

Bill Schneider conferred with his engineers about the cracked beams between the rocket stages. They convinced him that the cracks would not place the astronauts' lives in danger. Stresses during the launch would tend to close the cracks instead of pulling them apart. The decision was made to blast off Jerry Carr, Ed Gibson and Bill Pogue on Friday, November 16.

"We are continuing our examination of the rocket," Schneider said, "but we feel that the vehicle has adequate margin for safe flight, so we are committing to launch on Friday."

The cracks did not worry the astronauts. "The people that work on our launch vehicle are professionals," mission commander Jerry Carr told reporters. "We're extremely confident that when we launch, we're going to have a good bird under us."

November 16 began as a clear, bright day. At 8:01 A.M., the first-stage engines flamed to life. The rocket lifted off the launch pad only 1/400-second late. "Bird watchers" at Cape Kennedy could follow the rocket for a long distance as it

Third Skylab crew was launched from Pad B, Launch Complex 39, at 9:01:23 a.m., (EST), Friday, November 16, 1973.

streaked across the blue sky. They saw the first stage fall away and a bright flash as the second-stage engine ignited. It was a spectacular sight.

At the Kennedy Launch Center and Houston's Mission Control Center, engineers tensely watched their instruments and hoped they were correct about the cracks.

"To be truthful, I felt scared to death," launch operations manager Walter Kapryan told me. "We committed to the launch with a great deal of confidence, but there is always a moment of doubt. I personally sweated out those first seventy-two seconds [when the rocket undergoes the greatest stress in passing through the atmosphere]."

Ten minutes after blast-off, the astronauts went into orbit. They would spend Thanksgiving, Christmas and New Year's in space on a mission destined to last 84 days.

"Glad to Be Here"

The Skylab space station was making its 2,684th revolution of the Earth when the astronauts left the launch pad. They chased the station for another 5 revolutions and caught up to it about 3:30 P.M. over the Pacific Ocean.

Carr slowed the speed of the Apollo spacecraft until it moved only 0.2 mph faster than the space station. Then he gently eased the spacecraft's nose into the docking port. The capture latches slammed shut, but they did not catch hold of the ring on the Skylab station.

Carr tried to dock a second time. He failed again.

The controllers decided that Carr brought the craft together too slowly. The commander agreed. "I think I just hit it too easy," he said. "Can you imagine that coming from a Marine?"

Controllers recommended that Carr try docking again; this time moving the spacecraft faster. He speeded up to 0.6 mph faster than the station and moved into the docking port again.

"We got capture, Houston," he told the ground at 4:01 P.M. "We've got a hard dock."

"Welcome aboard," the control center replied.

"Glad to be here," Carr said with relief. "Great to be home."

The astronauts wanted to go aboard the station right away, but controllers recommended that they stay in the Apollo ship. Physicians felt that moving from the cramped spacecraft into the roomy station right away had something to do with making Bean's crew motion sick. The doctors decided that the crew should have more time to adjust to weightlessness before they started moving around a lot and doing hard work. They hoped to prevent sickness by having the astronauts take antimotion pills and spending their first night in the spacecraft.

It didn't work completely. Bill Pogue, the Air Force lieutenant colonel who had been an acrobatic pilot, felt so queasy that night he couldn't eat dinner.

Astronaut William R. Pogue (right), Skylab pilot, assists Gerald P. Carr, commander, during training and simulations in the Orbital Workshop trainer at the Johnson Space Center. Astronaut Edward G. Gibson (not pictured) is the science pilot.

Later, he vomited. About 2:20 A.M. the astronauts discussed what to do about this. They did not know that a tape recorder was running while they talked.

"We won't mention the barf," Carr said, "we'll just throw it down the trash airlock."

When the tape was played on the ground next evening, controllers were surprised and disappointed. Doctors needed all information about how the crew felt and what was eliminated from their bodies for the medical experiments. Also, controllers thought the astronauts should be absolutely truthful with them because they had to work together closely. The success of Skylab depended on close and complete cooperation.

Rear Admiral Alan B. Shepard, the first American in space and chief of the astronaut office at the time, scolded the crew. He told them that they had made "a fairly serious error in judgment."

Carr replied: "I agree with you. It was a dumb decision."

Fortunately, they did not throw the vomit bag away. The material was frozen and packed for return to Earth.

Next day, Carr and Pogue switched jobs. Pogue worked in the Apollo craft where he didn't have to move his head or the rest of his body too much. Carr and Gibson unloaded Apollo and got the spacehouse back in operation again. Pogue ate his lunch, and by Saturday night was working as hard as the other crewmen.

This was the emblem for the third manned Skylab mission. The symbols in the patch refer to the three major areas of investigation proposed in the mission: medical, scientific and technological.

A Surprise Welcome

A surprise awaited the astronauts when they moved into Skylab. They saw what looked like three men there. One sat on the exercise bike, another laid in the lower body negative pressure cylinder, and a third assumed an awkward position on the john. The last crew had stuffed three pairs of gold-colored coveralls with extra clothing and left these dummies to welcome Carr, Pogue, and Gibson.

"Hey, did you find enough food up there for six?" asked an astronaut on duty in the control center.

"The other three don't eat much," replied Gibson.

"They are also very quiet," added Carr.

Once they got settled, the astronauts had an important repair job to do. During the second Skylab mission, a refrigeration system had sprung a leak. This cooling system was different from the one that gave Conrad's crew a temporary problem as they were leaving Skylab. The present balky system cooled water and electronic equipment such as batteries. Bean's crew shut down this system and turned on a secondary or backup one. The secondary system also sprang a leak, but the leak remained slow enough to allow the system to keep working for weeks.

In both systems a material called Coolanol absorbed the excess heat and carried it away. When most of the Coolanol leaked away, the system would no longer do its job of cooling. Unless one of the cooling systems could be recharged with Coolanol, spoiled food and burned-out electronics might cut the mission short. Also, water cooled by this equipment was pumped into spacesuits to keep the astronauts from getting overheated during spacewalks. If water cooling did not work, an air cooling system could be used. But air-cooling was not as good, and it limited outside activity to about four hours. The astronauts figured they needed to spend five and a half hours outside on their first spacewalk.

On Monday, November 19, Pogue, recovered from his motion sickness, went to work on one of the cooling systems. Engineers compared the job with recharging a refrigerator in a home. Carr's crew had brought an 81-pound repair kit with them for the task, which was supposed to take about two hours. However, the equipment that checked for leaks in the recharging kit leaked itself. After Pogue solved this problem, he punched a hole in the cooling line, forced in refrigerant under pressure and sealed off the hole. The whole operation took four hours instead of two. When Pogue finished, both systems still leaked, and the leaks could not be found. But the crew could keep charging up both systems. They could make one long spacewalk on Thanksgiving Day instead of two on separate days. And the mission would not be cut short by problems such as overheated batteries.

Doing Upside-down Repairs

On Thanksgiving Day, Bill Pogue and Ed Gibson stepped outside the station. It was almost noon, and many people miles below them in the United States were getting ready for turkey dinners. The astronauts took photographs of dust and particles suspended in the Earth's atmosphere at high altitudes. Gibson changed film in the six camera telescopes. Then the two spacemen installed a number of detectors to measure dust, electrically charged particles, and cosmic rays near Earth. Engineers wanted to determine if these would harm windows, telescope lenses, and the insulating paint covering space vehicles on long missions. Then came the biggest job of all—repairing an antenna located on the bottom of the station.

The 4-foot-diameter, dish-shaped antenna belonged to an instrument used to measure the roughness and temperature of land and water surfaces under Skylab. The dish moved in a circle, picking up radiation emitted from the ground and radar signals bounced off the surface from the space station. It covered a ground area about seven miles in diameter. The antenna had begun moving erratically in September, during the previous mission. Gibson's and Pogue's assignment was to see what had gone wrong and to repair the problem.

One big difficulty involved finding a place to stand. There were no hand-holds or foot restraints between the exit hatch and the bottom of the station. Bill Pogue, a safety line hooked to his chest, worked his way hand over hand along a vent pipe leading to the antenna. He installed two foot restraints on a strut supporting the telescope mount.

Ed Gibson followed Pogue down the pipe with a tool kit and put his feet into the restraints. He was now standing with his head pointing down to Earth. Pogue laid across the bottom of the rounded workshop. Gibson held on to his shoulders, pushing and pulling him into position so that Bill could work on the antenna.

They worked like this for almost three hours. Ground controllers kept hearing one or the other say, "Hey, I floated out of position."

Removing six screws turned out to be the toughest part of the job. Astronauts and controllers thought one of them never would come out. Fortunately, the refrigeration system Pogue had repaired worked well and kept them cool. When the job finally was finished, the antenna worked about 75 percent as well as it did before the problem occurred. Bill and Ed spent a total of 6 hours and 31 minutes outside. This equaled the record spacewalk of Bean's crew on August 6, but it had been a much more difficult operation.

Miss Kathy Jackson, high school student from Houston, Texas, discusses with Arthur White (left) and Dr. Robert Allen her experiment, which was performed by astronauts in Earth orbit. The experiment tested eye-hand coordination during flight.

The antenna could now be used to do things such as measuring snow and ice cover—necessary for flood prediction when spring melting started. "We managed to save another major experiment," Neil Hutchinson said after the spacewalk. "We have about fifty scientists here with big grins on their faces. You just have to give the crew a great big star."

Astronaut Work for High School Students

Carr's crew performed some of the experiments designed by high school students, as had the two previous crews. On November 25, for example, the three astronauts tested their eye-hand coordination to see if it had been affected by weightlessness. Kathy Jackson, a student at Clear Creek High School, near Houston, Texas, suggested the experiment. The astronauts had to follow a certain sequence of inserting a thin stylus into 119 small holes arranged in a maze. They did this before the mission, three times during the mission and after splashdown. Kathy compared their performances to discover if a long period of weightlessness impairs an astronaut's ability to do fine, skillful work, such as assembling delicate instruments. The results showed no decrease in the astronauts' coordination after 84 days in space.

Joel E. Wordekemper and Donald W. Schlack designed experiments to determine if plants grow differently in space than they do on Earth. Wordekemper of Central Catholic High School in West Point, Nebraska, and Schlack of Downey High in Downey, California, used a box containing eight groups of rice seeds. They planted the seeds in agar—a jellylike nutrient extracted from seaweed. The seeds grew and sprouted in much the same way they did on the ground. The boys also wanted to see if the tendency of plants to grow toward light can substitute for their tendency to grow upward when gravity is absent. "At first we thought it might not because the growth seemed to be in a direction away from the light," said Henry Floyd, manager of the student program. "But after growing some distance in shadow, the plants did change direction and grow toward the light."

In another experiment, Robert L. Staehle of Harley School, Rochester, New York, determined with the help of scientists that bacteria grow in space much as they do on the ground. The bacteria Staehle used do not cause disease. Since the growth patterns of disease-producing bacteria are similar, however, scientists learned something about growth of microbes which might be harmful to astronauts.

These experiments with plants and bacteria provided the first available information on the possibility of using plants and bacteria on long space journeys, perhaps to distant planets. On such trips, plants could serve as a food source. They could remove carbon dioxide from the air and replace it with oxygen. Finally, they might provide a means of recycling human waste by using it as fertilizer.

Searching for a New Planet

Terry C. Quist, then a student at Thomas Jefferson High in San Antonio, Texas, thought up one of the most successful experiments. He wanted to determine the amount of neutrons that enter a spacecraft orbiting near Earth. Neutrons are part of the core, or nucleus, of all atoms. Scientists bombard other materials with neutrons to make them radioactive. Terry Quist wanted to find out how many neutrons might be flying around in a spacecraft and if they posed any danger. In addition, he hoped to learn what portion of the neutrons came directly from the Sun, how many were reflected from the Earth, and what amount originated when cosmic rays from outer space collided with atoms near Earth.

Terry faced the problem that these neutrons have such high energies they whiz right through a spacecraft without slowing down enough to be detected. Water slows down neutrons, so he hit upon the idea of using water storage tanks aboard Skylab to reduce the energy of the neutrons. With the help of his teacher-advisor, Mr. Michael Stewart, and some NASA engineers, Terry designed neutron detectors. Eleven of these were mounted in the Skylab workshop. Examining detectors brought back to Earth, scientists were amazed to discover about twice as many neutrons as expected. Tracks left in the detectors also showed that the particles had more energy than suspected. Thus, scientists learned something new and basic about what goes on in space and what kind of radiation a spacecraft is exposed to when orbiting close to Earth.

Troy A. Crites of Kent, Washington, thought he might be able to discover a way to predict volcanic eruptions from information obtained aboard Skylab. Instruments in the Earth Resources Experiment Package measured the amount of heat given off by different objects on the surface. Scientists reasoned that volcanic eruptions would be preceded by a build-up in heat or infrared radiation. Troy figured this buildup could be sensed by a spacecraft passing overhead. He wanted to record patterns of heat radiation from active volcanoes. Then he would compare these with measurements on the ground and how long it took a volcano to erupt once a certain amount of heat had built up. From this, he thought it would be possible to discover when energy accumulates to the point where it will be released in a dangerous outburst.

Astronauts on all three missions made measurements of either Cerro Negro in Nicaragua or Mt. Etna in Sicily, both active volcanoes. Geologists hope this will be one step toward establishing an eruption forecast system, which would allow people to be evacuated in time to avoid loss of life, injury, and damage to property.

Donald C. Bochsler of Silverton, Oregon, had an ambitious project. He wanted to settle the question of whether an undiscovered planet orbits so close to the Sun that no one has been able to spot it. Some astronomers suspect that a small planet orbits between Mercury and the Sun, and one man went so far as to name it "Vulcan." Donald figured to use the coronagraph in the telescope mount, which blocks out the brightest part of the Sun. Photographs taken through the telescope on this instrument show the Sun's outer atmosphere and black space nearby. If Vulcan does exist, it might show up in these photos.

"The possible discovery of another planet made this a very exciting experiment," said John MacLeod, a student advisor at Johnson Space Center. "The person who named it Vulcan did not prove its existence. If the student became the first person to actually see the planet in photographs, he would get to name it." But, so far, no new planet has been found.

Of the nineteen student experiments that were included in the Skylab flight plan, five were not considered successful. However, all twenty-five of those who won the national competition received data from Skylab. They could examine the data and write up their conclusions in a scientific paper.

"Part of the purpose of the student program was to help get the taxpaying public more interested in space," admits MacLeod. "We also wanted to give students the opportunity to participate in something far beyond the realm of ordinary high school activities. We would have been successful in these goals even if no good results had been obtained. But the scientific payoff has been far more than we thought it would be. The student program turned out to be successful beyond our fondest hopes."

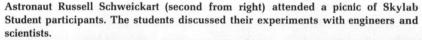

Astronaut Russell Schweickart (second from right) attended a picnic of Skylab Student participants. The students discussed their experiments with engineers and scientists.

Chapter 12

The Comet That Wasn't There

The six-and-a-half-hour spacewalk on Thanksgiving Day had been an exhausting one. After it was over, the astronauts ate supper, asked the ground about some football scores and then went to bed about 11 P.M.

As they slept peacefully, problems continued aboard trouble-plagued Skylab. A 400-pound wheel, spinning 9,000 times a minute, began heating to dangerous levels. Three such wheels, spun by motors, formed part of a gyro system that controlled the position of the 100-ton-space-station-Apollo spacecraft combination. This system kept it flying straight and level, or provided a force to change its position.

Instruments on the gyro wheel sent signals to Mission Control Center showing sharply rising temperatures in one of the wheel bearings. The heat went from a normal 80 degrees up to 180 degrees. The graveyard shift of controllers decided to shut down the gyro at 2:45 in the morning.

The crew awoke just before 6 A.M., and astronaut Hank Hartsfield told them what had happened. "Last night, we lost gyro number one," he said. "It looks like a bearing seized up on us, so we're down to two gyros."

The space station could operate normally on two gyros, so the astronauts didn't need to take any immediate action. They decided the best thing to do was to get an extra half hour of sleep. Ground controllers agreed. They took care of setting up Skylab to run on two instead of three gyros.

The gyros work in conjunction with thrusting rockets to control and maneuver the space station. Maneuvering commands come through a computer from the astronauts, from ground controllers, and from small sensor gyros. You remember that Jack Lousma replaced a six-pack of these small gyros on a spacewalk in August. They sense unwanted movement of the station and signal the computer. The computer automatically sends commands to correct the movement. Also, the astronauts could operate the gyro system with switches on a control console in the docking adapter. Or controllers could run the wheels and thrusters by sending commands up from the ground.

All this could be done with two as well as three gyros. But if a second gyro failed, then the astronauts would be in what controllers called "a come home situation." One gyro could hold the station steady with the help of thrusters on the Apollo spacecraft. But in this situation, the maneuvers needed for Earth surveys and other experiments could not be made. This would make Skylab almost useless, so that the astronauts would be brought home.

With two gyros these maneuvers could be made, but more thruster fuel would have to be used. "It'll take a little more gas to get there now," flight controller Phil Shaffer told reporters on the morning of November 23. "And it will take more time to do the maneuvers."

Kohoutek Is Coming

One type of maneuver which concerned everyone involved moving the station to get the best view of the "comet of the century." On March 7, 1973, a Czech astronomer named Lubos Kohoutek discovered a new comet some 440 million miles from the Sun and 350 million miles from Earth. He spotted it on photographs taken through a telescope at the Hamburg Observatory in West Germany. Kohoutek actually was searching for asteroids when he happened to see the comet. "It was really an accident," he told me later.

No one knows where comet Kohoutek, or any comet, began its existence. Lubos Kohoutek and other scientists believe comets consist of frozen balls of ice, dust, and gas left over from the creation of the Sun and planets. This would mean they are 4 to 5 billion years old. The Czech astronomer says that on the farthest edge of our solar system, 4 to 5 trillion miles from Earth, exists "large clouds of comets." Occasionally one of them passes near a star, and the gravitational tug between the two sends the comet on a slow journey toward the Sun. Kohoutek thinks that about 2 million years ago this happened to the comet named after him.

A few years before he spotted it, Kohoutek believes that the large outer planets—Neptune, Uranus, Saturn, and Jupiter—changed the comet's path again. They bent its orbit so that it would pass about 13 million miles from the Sun and 75 million miles from Earth.

When Kohoutek first spotted the comet, it appeared very bright. As these frozen balls of ice come closer to the Sun, heat vaporizes, or boils away, gas and

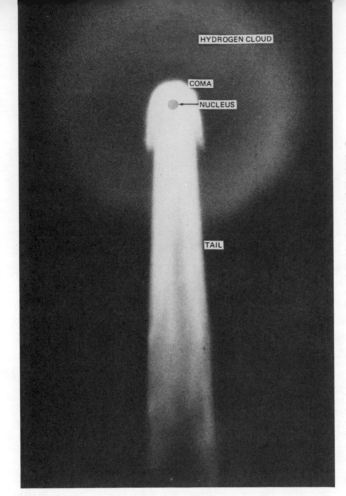

HYDROGEN CLOUD

COMA

NUCLEUS

TAIL

Comets are regarded as samples of primordial material from which the planets were formed billions of years ago. The most permanent feature of a comet, the nucleus, is believed to be a sort of dirty ice ball, consisting of frozen gases and dust particles. Dust particles liberated from the comet nucleus form the dust and plasma tails, which can extend up to 100 million kilometers (60 million miles).

dust. This produces a glowing head, which may be thousands of miles in diameter, and a fiery tail stretching out for millions of miles. Although the ball of ice at its core probably reached no more than 20 miles across, astronomers predicted Kohoutek would be as bright as a half moon and have a tail 50 million miles long.

By October, newspapers called it the "comet of the century"—the most dazzling sight in the sky for the past 100 years. Since it would be closest to the Sun, and therefore at its brightest, near Christmas Day, a NASA press release compared it to "the Star of Bethlehem." "It is expected to be more spectacular than the famed Halley's comet, which last appeared in 1910, and as bright as the full Moon," said the release.

Never before had a comet been spotted so far away on its approach to the Sun. Usually, astronomers didn't spot these mysterious invaders until only two months remained to prepare for them. But from March until the November launch date of Skylab, scientists had almost nine months to ready their instruments. And astronauts would have a heavenly front row seat when Kohoutek made a Christmastime swing around the Sun and headed for the outer reaches of the solar system again.

"It's the first time that we've ever had a complex observatory in space when a comet was due to come by," noted William C. Snoddy, a NASA scientist who worked on the Kohoutek program. "It may never happen again." While observers on the ground could see the comet only once a day from late November to mid-January, astronauts would be in a position to observe it thirteen times a day on each orbit of Earth.

No one knew what kinds of gas and dust make up a comet. The ultraviolet radiation given off by the heated materials would be important clues for determining this. But Earth's thick atmosphere scattered and absorbed this radiation so that it never reached instruments on the ground. Besides ultraviolet and x-ray instruments already aboard Skylab, scientists readied another ultraviolet camera for Carr's team to carry up with them. This camera happened to be a backup for one that astronauts used on the Moon's surface during the *Apollo* 16 Moonwalk in April 1972. On Christmas Day and on December 29, the astronauts would attach it to struts supporting the telescope mount and take pictures of the comet of the century.

"The idea that we get a ringside seat for such an important event—that we will see what no others have ever seen—is thrilling to me," Carr commented.

In addition to Skylab, scientists planned to study the comet with planetary probes and satellites already in space, high-flying aircraft, instrumented rockets, a vast network of ground observatories, and even telescopes carried aloft by balloons.

A Spacewalk for Christmas

On Skylab, the plan was to make observations of the comet almost every day from November 23 until it got close to the Sun. This would be done with instruments pointed out of an airlock. As the comet approached the Sun, observations would switch to the telescope mount instruments and to spacewalks. All this required maneuvering the station. Engineers on the ground calculated that enough fuel remained to do this with one gyro not working. They could also use rockets and some fuel aboard the Apollo spacecraft for this purpose.

On December 9, Carr reported that Kohoutek "is getting easier and easier to see. The tail is quite visible and quite long."

The tail develops when gas from the Sun, called solar wind, blows gas and dust from the head of the comet. Carr noticed that Kohoutek's tail started to fork or split. Solar winds affect dust and gas in a different way. The heavier dust forms a fuzzy, slightly curved, whitish tail. The lighter gas forms a separate, straighter tail with a bluish color.

On Christmas Eve, the astronauts reported that the head had grown much larger and brighter. The split tail fanned out and changed from blue and white to tints of yellow, orange, and red. Kohoutek moved toward the Sun at speeds of 200,000 mph.

Next day, Jerry Carr and Bill Pogue stepped outside for a better look. With the comet this close to the Sun, and material rapidly boiling off and glowing, the view should be spectacular. But the astronauts would be the only ones to see it. Kohoutek was now too close to the Sun's glare to be spotted from Earth.

About 10:50 A.M., the astronauts began setting up three instruments on the telescope mount supports. One photographed x-rays and ultraviolet radiation from the Sun; another served as a coronagraph to block out the dazzling solar disk so Kohoutek would be visible. The third was the special camera for photographing the comet in ultraviolet light. Ed Gibson, inside, maneuvered the space station so the solar panels blocked out the Sun. This would make it easier for Carr and Pogue to see Kohoutek.

The astronauts tried hard, but they could not spot the comet. They had to be satisfied with pointing the camera to where it should be in the sky. As it turned out, they got thirty-three pictures of Kohoutek.

Next, Carr went out to change film in the telescope mount cameras. He also had to fix a jammed filter which had blocked the lens of one camera since November 26. This latter job took much longer than everyone thought it would. By the time Carr succeeded in clearing the obstruction from the lens, the scheduled five-and-a-half-hour spacewalk had stretched to seven hours. It turned out to be the longest spacewalk on all three Skylab missions. Carr's crew now had more time outside the space station than either of the previous two crews.

"Be it ever so humble, there's no place like home," Carr said when he finally got back inside.

This photograph of the comet Kohoutek was taken at the Catalina Observatory by members of the lunar and planetary photographic team from the University of Arizona.

A Thrilling View

Carr went outside again at 11:30 A.M. on December 29, with Ed Gibson. Kohoutek had looped around the Sun on December 27–28 and was now moving back out toward deep space.

"Hey, I see the comet," yelled the scientist-astronaut as Skylab orbited into the darkness over Europe.

"Holy cow, yes!" replied Carr. "Beautiful! The tail has fanned out. It's wide, broad, and colored yellow and orange like a flame."

Since solar wind blows dust and gas away from the comet's head, the tail always points away from the Sun. Now that Kohoutek had changed direction, the tail shot out in front instead of streaming behind. In addition to this long tail, the astronauts spotted a short tail, pointed toward the Sun and stretched out behind Kohoutek.

"There's a spike going straight toward the Sun," said Gibson. "That [comet] is one of the most beautiful creations I've ever seen. It's very graceful."

At first, astronomers were puzzled by the spike, or antitail, as some call it. Later, they figured that the spike comprised heavier dust particles which hadn't been blown around to the other side of the comet by solar wind. The particles were large enough to resist the forces from the Sun for a longer time. Not all astronomers believed that such antitails could exist. Thus, the astronauts obtained the first direct proof of this, and the first evidence that such large particles are present in comets.

Once around the Sun and heading for its closest approach to Earth on January 15, Kohoutek, astronomers predicted, would be a spectacular sight. William Liller, professor of astronomy at Harvard University, said that the tail could be as long as 20 million miles on January 15. Planetariums advised people that the comet would be clearly visible without binoculars or telescopes. All they had to do was look for it in the western sky just after sunset.

This didn't turn out to be the case. Instead of becoming brilliant after its searing encounter with the Sun, Kohoutek faded rapidly. On January 4, the astronauts reported it was so dim they could hardly see it. The same day, Stephen P. Maran, head of NASA's Operation Kohoutek, spotted the comet without binoculars from a research plane flying high over Pennsylvania. He calculated that its yellow tail streamed 6 million miles across the sky. But Kohoutek never got as bright as predicted. It was no Star of Bethlehem. Dr. Kohoutek himself said the comet turned out to be dimmer than the planet Venus, and its tail never got longer than about 8 million miles.

During the comet's passage within view of Earth, few non-astronomers saw it with the naked eye. Those who knew where to look, and had a pair of 7x50 binoculars, easily could see a fuzzy tail. But they could not detect any movement. "For the general public," said Maran, "Kohoutek was the comet that wasn't there."

What went wrong? No one really knows, but, as usual, scientists have some theories. One is that a chemical reaction produced a sticky "glue" that coated the surface and prevented the dust and gas from escaping. Other astronomers think that the level of activity on the Sun was too low to set much gas and dust shining brightly. According to one of the best ideas, Kohoutek seemed so bright when first spotted because it was surrounded with small chunks of frozen ice that reflect sunlight very well. When the comet got closer, the chunks melted away. Yet another theory holds that dust formed a bright halo around the comet when it was still far away, leading everyone to calculate that it would be much brighter as it heated up. But Kohoutek contained only a relatively small amount of dust, and this turned the comet of the century into the comet that couldn't.

However, astronomers were not as disappointed with Kohoutek as the public. Maran and others called it "a large, exciting, relatively bright comet." Seen through telescopes at mountain-top observatories in such places as Hawaii, Arizona and New Mexico, it was a superb sight in mid-January, Maran noted.

Scientists obtained information that cleared up some mysteries about comets and provided intriguing clues to the solution of others. The antitail question was settled. Astronomers became virtually sure that the "dirty snowball" theory is correct. That is, that comets are icy spheres, or heavenly snowballs, packed with the type of dust found between the stars. They learned that this dust and ice is more complex and exotic than they thought. The data revealed substances never known to be present before in comets—things like hydrogen cyanide. These same substances occur in parts of the Milky Way where astronomers believe new stars and planets are forming.

"The findings imply," Snoddy points out, "that comets form in the extreme region of our solar system, beyond Pluto, and are very likely made of the primordial material out of which our Sun, Earth and planets originally formed." Thus, these beautiful and mysterious objects must contain clues to the origin of our solar system. To discover such clues, NASA hopes to send unmanned space probes to a comet called Encke in 1980 and to the famous Halley's comet when it returns in 1986. What they learned from Kohoutek will help them plan these missions.

Welding in Space

On Saturday morning, January 5, Jerry Carr went into the docking adapter to start an electric furnace. This furnace made up part of a facility—called a materials processing facility—that someday may be remembered as the beginning of factories and repair shops in space. The astronauts used it to do the first U.S. experiments in welding, metal working and growing electronic crystals. These crystals go into circuits that operate television sets, calculators, computers and most other electronic devices.

Years before the first Skylab flight, engineers began to study how conditions in space could be used to make valuable products that could not be made on the ground. Weightlessness, they figured, would make it possible to mix metals and chemicals that wouldn't mix on Earth because of gravity. Melting substances inside containers always caused the finished product to pick up impurities from the walls. In space, crystals could be made while floating free of walls that caused such problems. These possibilities led to the building of a combination vacuum chamber and electric furnace and its installation aboard Skylab.

Engineers also wondered if welding was possible in weightlessness. Would it be possible to construct and repair large vehicles and stations in space? The Soviets had tried welding and cutting metals in space aboard the *Soyuz 6* mission in October 1969. Therefore, the Russians were ahead of the United States in this work at the start of Skylab. However, as in other aspects of space operations, the Americans caught up, then passed them.

Pete Conrad's crew began the first experiments in June. They didn't actually don welder's gloves and masks and join metals together. Operations were done

Astronaut Charles Conrad, Jr., works with the possibility of building factories in space.

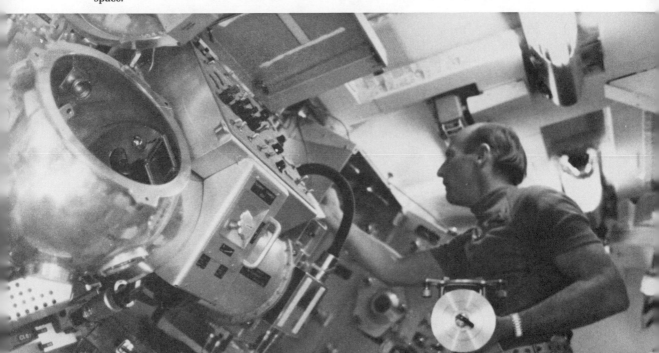

automatically in the vacuum chamber. The astronauts put the experiments in, turned on the furnace, and movie cameras recorded what happened. Then they removed the finished samples and packed them for return to Earth.

Inside the vacuum chamber, an electronic beam melted metals the way they would be melted in welding. Experts feared that, without gravity, the molten fluid would float out from between the edges of the metals to be joined. But the experiment proved it would stay in place, solidify, and hold the pieces together. When they examined the samples on the ground, engineers concluded that space welds are as good as, or better than, welds made on Earth.

Conrad's crew also did experiments to determine if nickel and stainless steel tubes could be joined by brazing. In brazing, a filler between the metal is melted and solidified instead of the metals themselves being melted. Again, engineers feared that the brazing filler would run out. This didn't happen. Metal joints with gaps as wide or wider than those that can be brazed on Earth were successfully joined together. Engineers inspecting the space braze joints on the ground called them "outstanding" and "much finer" than similar work on Earth.

This work showed definitely that repair and construction work in space is possible. It paved the way for astroworkers to repair large spacecraft without bringing them back to Earth, and to assemble buildings in space that would be too large to launch on a single rocket.

Making Unearthly Crystals

Bean's crew did eleven out-of-this-world materials experiments. They turned out so successfully that Carr's crew repeated seven of them to get more data. The results prove valuable products can be made commercially in space, according to Dr. William R. Lucas, director of the Marshall Space Flight Center.

The experiment Jerry Carr did on January 5 involved trying to grow crystals that were unearthly pure and perfect. Engineers cut such crystals into thin wafers or chips. Transistors, diodes, and other electronic parts placed on the chips form circuits that do things such as run a computer or an electric power switching system. A single wafer, less than an inch in diameter, can hold about 750,000 of these electronic parts. The better the wafers or crystals, the smaller, more efficient, lighter, and less expensive will be the television set or computer made from them.

The most important ingredients in these crystals are impurities called dopants. On Earth it is almost impossible to get the dopants uniformly distributed through the crystal. Before the crystal hardens from molten liquid, gravity causes the hottest parts of the liquid to move upward and the cooler parts to sink. Such movement, or *convection,* prevents dopants from spreading evenly through the crystal. Engineers guessed that such movement would not occur in weightlessness, therefore they could make crystals that would be much better than those made on the ground.

Sure enough, when engineers studied space-grown crystals they found a uniform distribution of dopants. Harry C. Gatos, the man in charge of one experiment, and a professor of metallurgy at the Massachusetts Institute of Technology, was extremely pleased. Achieving such uniformity, he said, will usher in a new era in which materials will be thousands of times stronger and better.

Another experiment run by both Bean's and Carr's crews produced crystals larger and more perfect than those grown on Earth. The experiment used a method which, on the ground, produces crystals too small to be practical. Gravity effects prevent larger crystals from forming. In space, it was expected that such crystals would reach a length of ⅛ inch. Actually, one crystal turned out to be ⅔ inch—6 times longer than expected. From this result, engineers may someday design methods for the commercial production of crystals in space.

Chemically pure and better-structured crystals were grown in other experiments by other methods, too. The results led engineers to conclude that huge, flat crystals could be grown that would do away with cutting and polishing into wafers. Cutting and polishing produce strained, contaminated chips and create a lot of waste. Pure, flat crystals could help solve the energy problem by reducing inefficiency in power transmission and making possible cheap, efficient solar cells to convert sunlight to electricity.

Substances never made on Earth came out of one experiment. Astronauts mixed materials that are completely unmixable on the ground because gravity causes them to separate like oil and water. The experiment produced an unearthly compound of gold and a hard, grayish-white metal called germanium. The alloy shows evidence of having no resistance to the flow of electricity when cooled to extremely low temperatures. Such *superconductors* can be made on Earth, too. But perhaps they could be made better in space. Either way superconductors could revolutionize present methods of moving electricity from the places where it is generated to places where it is used.

Everyone involved was surprised and astonished at how well the experiments worked. However, it will still require much study, planning, time, and more experiments before factories in space become possible. Just the very first steps have been taken. More materials experiments were done on the joint U.S.-Russian flight in 1975. Hundreds more are planned for missions aboard large spacecraft in the early 1980s. Perhaps in 1985 or 1990, NASA, in conjunction with private industry, may establish the first pilot plant in space to produce drugs or electronic crystals.

James H. Bredt, NASA's manager of space processing applications, points out that a lot of new basic knowledge about materials came out of the Skylab experiments. "Anytime you have new knowledge like this, somebody finds a use for it," he says. "The knowledge will lead to inventions and products that we cannot even think of today."

Discovering Earth from Space

Jerry Carr, Ed Gibson and Bill Pogue were the second crew of astronauts to spend Christmas in space. Five years before them, in 1968, Frank Borman, James Lovell, and William Anders became the first men to circle the Moon, and they did this on Christmas Eve. They thrilled the world by reading from the Bible as they sent back the first close-up pictures of the Moon's surface taken by man.

The Skylab astronauts had an exciting Christmas too—walking in space and trying to photograph a comet as it neared the Sun. Carr's crew also had fun making a Christmas tree. Long, thin racks, which held cans of food, formed the branches. The empty red, white, and blue cans themselves served as ornaments. They topped this off with tinsel of green, red, silver, and gray tape.

On Christmas Eve, the astronauts' families visited the Mission Control Center and saw television pictures of the tree. In fact, they saw two trees. Ground crews had hidden presents in the Apollo ship under equipment needed for the spacewalk. One box held a 3-foot-tall tree made out of fireproof green cloth, trimmed with flameproof silver and gold ornaments.

The astronauts exchanged greetings with their families and the ground teams as they flew over Texas on Christmas Eve. Everyone experienced the kind of good feeling that makes you want to laugh and cry at the same time. The families missed their men, but they were able to joke and laugh. Bill Pogue's eighteen-year-old daughter, Lana, sent him this message:

Jingle bells, jingle bells.
Fa-la, fa-la, fa-la.
We opened all your Christmas presents.
Ha, ha, ha.

Breaking All Records

Skylab astronauts celebrated Christmas on their fortieth day in space. By January 11, they had been in orbit for 56 days. Carr, Gibson, and Pogue exercised and slept more than previous crews, and doctors judged them to be in better physical shape. Their bodies seemed to have adapted to weightlessness by the fortieth day. No important medical changes had taken place since then.

During the first month, doctors had worried about the crew. They seemed to get tired easily and to work slower than Bean's crew. They made fairly serious mistakes. Besides the vomit incident, Pogue forgot to put filters on a camera used to take survey pictures of the Earth. Valuable data had been lost because of this. The crew experienced trouble with new medical experiments on their mission. Gibson blamed this on not enough training for the tasks.

All this was behind them by January. Bill Schneider admitted Carr, Gibson, and Pogue had not done as much as Bean's crew. But he said: "Their performance has been up to what we expected of them. [Bean's crew performed beyond expectations.] "We've had a successful and productive mission so far."

At launch, NASA was committed to a 60-day mission. This was to be extended to 84 days if the crew stayed healthy and the spacecraft remained in good condition. "Both crew and spacecraft are in good shape," Schneider reported on January 11. "We have enough consumables [food, water, fuel, oxygen, etc.] and enough work to do to make an extension of the mission desirable. And so, we give a GO for a longer mission."

Three days later, at 8:10 P.M. on January 14, 1974, Carr's crew passed the 59-day, 11-hour record set by Bean, Garriott, and Lousma. They had circled the Earth 858 times and traveled more than 24 million miles since November 16. Every minute more they spent in orbit would increase the length of man's longest spaceflight.

On January 25, they broke Al Bean's record of 69 days in space, his combined time on Skylab and a flight to the Moon.

Controllers felt that only one thing could bring Carr's crew down before the planned 84 days. Since December 10, a second gyro had been giving trouble. It sped up and slowed down, heated up, then cooled down—just the way the first gyro acted before it had to be shut off. If the second gyro failed, the astronauts would take a spacewalk to get film in the telescope mount, then return home. In case both remaining gyros failed, or another major problem came up, a rescue rocket could be launched in only nine days.

The gyros had taken a toll of one of the most important group of experiments aboard—the Earth resources survey. Skylab had to be maneuvered so that the cameras and other instruments faced their targets on Earth. Then clear weather was needed for the best view of the targets. Problems with the gyros made maneuvering difficult, and the weather had been cloudy.

A Special View

Skylab's 270-mile altitude and special orbit gave the astronauts and their instruments a unique look at Earth. Engineers planned the orbit so that the space station flew over 75 percent of the Earth's land and oceans, including those places where 90 percent of the world's people live and 80 percent of the world's food is grown. It passed over all parts of the United States, except Alaska, during daylight hours. It went over all of Africa, Australia, and Japan, most of Europe and South America, and much of Asia.

No manned American spaceship had ever flown over so many countries. The Apollo, Gemini, and Mercury spacecraft never passed over places farther north than Los Angeles or farther south than Buenos Aires, Argentina. They never flew over any part of Europe. But Skylab's flight path took it as far north as Winnipeg, Canada, and Frankfurt, Germany, and as far south as the southern tip of South America. It was the first time astronauts looked down on such major cities as San Francisco, Chicago, New York and Washington, D.C.

Skylab went around the world once about every ninety minutes. The Earth moved under the space station in such a way that it came over the same place about every five days. That meant Skylab could check the same area every five days over a period of nine months to see what changes occurred.

The astronauts saw fabulous sights out of their 18-inch porthole. But this wasn't a sight-seeing trip, and scientists wanted more information about Earth than the eye could supply. Therefore, the astronauts' senses were extended with six special instruments known as the Earth Resources Experimental Package (EREP). Located in the docking adapter, EREP consisted of special cameras, radar, and instruments that recorded radiation not visible to the eye. Just as the Sun gives off invisible ultraviolet and x-rays, so every object on Earth radiates heat, or infrared rays. An object can be identified by the pattern of heat it gives off as easily as if you saw it by eye. Combining both visible and infrared views of a scene on Earth tells a lot more about what is there than a color picture alone.

EREP included instruments to measure microwaves emitted by land and water. These consist of noiselike signals whose character depends on the temperature of the surface giving them off. Microwaves pass easily through rain, snow, fog, smoke, and clouds. Therefore, microwave instruments gave the astronauts a way to look through the clouds which block visible light and infrared rays.

With these Earth-watching devices man could see many things he could not see with his eyes alone. Five instruments, mounted together in the bottom of the

A vertical view of the Strait of Gibraltar area photographed from the Skylab space station in Earth orbit. The largest land mass is Spain, appearing in the northern half of the picture. A small portion of Portugal appears in the northwest corner of the photo.

docking adapter, were positioned to look at the same area of Earth. One instrument consisted of six cameras which took photos of a region 100 miles on a side. One click of the cameras produced 6 photos of a 10,000-square-mile portion of the Earth. EREP also contained the largest telescope to be flown in a manned spacecraft. It had a 24-inch mirror system which scanned across a 26-mile-wide path under Skylab, recording visible light and infrared radiation. The station orbited over Earth at a speed of about 4 miles a second, so this telescope surveyed a 104-mile swatch of land or water each second—6,240 miles each minute. A sixth instrument—a camera—was mounted in one of the two airlocks in the station workshop. It took photos of objects as small as 35 feet from 270 miles in space.

Prospecting from Space

The purpose of Skylab's EREP program was to test how useful these instruments would be for exploring and mapping the Earth, and for keeping an inventory of its resources. Skylab was far more efficient in this than an aircraft. To cover the same ground so often would require squadrons of planes and fleets of ships. Some of the remote places in the world could only be surveyed from a spacecraft. While the instruments were being tested, they obtained practical information about such things as crops and forests, insect plagues, the presence of oil and metal ores, volcanoes, earthquakes, hurricanes, storms, snow and ice, wave heights, air and water pollution, areas of drought, and the growth of cities.

Such information from space had proved its value. Ex-astronaut Walter Cunningham told me about an Egyptian geologist who was amazed by a photo that one astronaut took. It showed a mineral deposit which he had worked on in Egypt. The startling thing was that the photo showed the mineral deposit to be four times larger than the geologist, or anyone else, had thought.

Photos from space had shown evidence of new copper deposits in Colorado and oil and tin deposits in Alaska. When prospectors examined photos taken by Conrad's crew, they discovered a possible copper lode in Nevada. Prof. M. L. Jensen of the University of Utah spotted a light-colored patch in the middle of some dark volcanic rocks near Ely, Nevada. He had flown over this area in a plane but never noticed this. Jensen suspected the whitish rocks consisted of limestone which might contain copper deposits. A check by ground teams confirmed this, and Jensen reported a rush to stake claims in the area.

Volcano and Hurricane Watching

On three Skylab missions, astronauts took over forty thousand pictures of Earth. Information not collected on film was recorded on magnetic tape, and the spacemen filled about forty-five miles of tape with data. Photos and tapes contained information about all the states, except Alaska, and about thirty-four foreign countries. "These results far surpassed what we thought would be accomplished," said Dr. Verl R. (Dick) Wilmarth, an EREP scientist at the Johnson Space Center.

One hundred thirty-seven scientists and their staffs in the United States and eighteen foreign countries studied this mass of data. They obtained information about such widely different things as walnut worm attacks on pecan groves in Texas, and possible sources of water to help starving people in drought-stricken parts of Africa.

Jerry Carr and his crew carried up and installed an instrument to detect new hot springs and volcanic areas. The hot water and steam in such places may be tapped as a source of energy. All the Skylab crews collected information about active volcanoes in Nicaragua, Mexico, and Sicily. This information might help

A striking example of a geologic fault in the Earth's surface is evident in this photograph of the Tien Shan Mountains of western China.

to reduce death and damage from eruptions, and to tame volcanic energy for heating homes and other useful purposes.

Carr, Pogue, and Gibson turned on EREP instruments over the remote Gran Chaco plain in Paraguay, Argentina, and Bolivia. Much of this area of swamps, jungles, and grassy plains had never been mapped. "They demonstrated that EREP sensors can provide us with the kind of data we need for mapping inaccessible areas without a major program of work on the ground," said Wilmarth.

Skylab crews spotted sources of air and water pollution from orbit. This proved that EREP sensors could be used on unmanned satellites to keep track of pollution. Some scientists used Skylab data to study strip mining and determine its effects on ecology. Others used different information to monitor changes that took place in thirteen U.S. cities since the 1970 census, and to keep track of rapidly expanding urban areas.

Weathermen got a special bonus when Skylab passed over a large storm or hurricane. In early January, Carr's crew repeatedly flew over one of the largest North Atlantic storms in a decade. "We obtained data on winds blowing seventy miles per hour and waves forty to fifty feet high," Wilmarth said. "This should enable us to determine the best kinds of sensors to use on weather satellites."

Conrad's crew photographed hurricane Ava off the west coast of Mexico on June 6, 1973. At the same time, airplanes with special instruments flew through the violent winds at various heights from 500 to 10,000 feet. The sensors recorded winds up to 150 mph and waves as high as a 4-story building. "This made Ava one of the most intense hurricanes recorded in that part of the Pacific," according to a NASA report. EREP instruments recorded wind speeds from less than 15 to 150 mph, cloud conditions from clear skies to towering hurricane clouds, and sea states from nearly calm to 45-foot waves. Said the report: "Equivalent coverage could only have been obtained by either hundreds of specially instrumented buoys and ships, or several fleets of aircraft."

The hurricane survey was one of a number of experiments in which observers in airplanes and on the ground collected information at the same time as the Skylab crew. This enabled engineers to check the EREP instruments against what actually occurred on the ground. For example, a field of corn suffering from blight would produce a certain pattern of visible, infrared, and microwave data. Once you verified what this pattern meant with observations on the ground, you could quickly check large areas for blight with a spacecraft.

Conrad's crew did an experiment like this on the first mission. With EREP instruments turned on, Skylab flew over the eastern part of the Gulf of Mexico, off Florida. Down below, 110 sports fishing boats and 4 research ships caught fish and collected samples of the water. Putting all this information together produced a pattern which shows when fishing should be good or bad. Fishermen spend most of their time looking for fish instead of fishing. Being able to check the ocean with a satellite that radios back locations where fish might be, will reduce the cost of catching fish. It is hoped that knowing where different kinds of fish are at various times of year will help prevent overfishing and the extinction of some of our favorite seafood.

Using the Eyes

Bean's crew added something new to Earth-watching. When they got ahead of schedule and asked for more work, ground teams gave them a list of things to watch for and photograph on Earth. These included ice that might get in the way of ships, heavy snow that might melt and cause spring floods, sand dune patterns, and pollution sources. Scientists decided to continue these visual observations on the last mission. They included them in the flight plan, and Carr's crew received special training from eighteen experts in different fields. "The purpose of this program," explained Wilmarth, "was to determine how useful a man is for getting data of direct interest in studying our Earth."

Dr. Robert E. Stevenson, an oceanographer with the Office of Naval Research, spotted huge whirls, or eddies, in the Yucatan Current, which flows northward along the east coast of Mexico. He saw these in photos taken by Conrad's crew. Stevenson then asked Carr, Pogue, and Gibson if they would look for these eddies and take more pictures of them.

"I had grave doubts we'd be able to see much in the way of ocean currents," Carr said. But a surprise awaited him.

"One day you are going along watching nice streaks of clouds," he told a press conference afterward. "Then all of a sudden you find great big empty places." Astronauts looked down through these holes in the clouds and there were the eddies.

These ocean whirls, forty to fifty miles across, mark places where cold water comes up from four hundred or more feet below the surface. Usually ocean water is warmer than the air above it. The water warms the air, causing it to rise. Condensation occurs and clouds form. But where water is colder than air, clouds do not form. Therefore, the eddies became easy for Carr and crew to spot.

Once they knew what to look for, the astronauts found eddies all over the world—off Africa, Australia, and both coasts of South America. Oceanographers knew about eddies, but they didn't know they occurred worldwide before Skylab. The chief of the Office of Naval Research called this "the biggest discovery in oceanography in the past ten years."

Astronauts found a new eddy near French Frigate Shoal in the northern part of the Hawaiian Island chain. The Navy had a fleet of ships in the area when Skylab flew over, but the sailors couldn't see the eddy although they sailed right through it. This proved the value of being able to get far above the Earth and look back at it.

The astronauts also discovered a new eddy off the east coast of New Zealand, where water from two different ocean currents comes together. Such upwellings of water bring to the surface mineral nutrients on which ocean plants feed. This produces "blooms" of microscopic floating plants. The plants become food for billions of tiny floating animals. Collectively known as *plankton,* the plants and animals make up a rich broth that nourishes larger sea animals. Creatures from shrimp to blue whales dine on plankton. Those that don't eat plankton eat plankton-eaters. The tiny animals and plants occur in

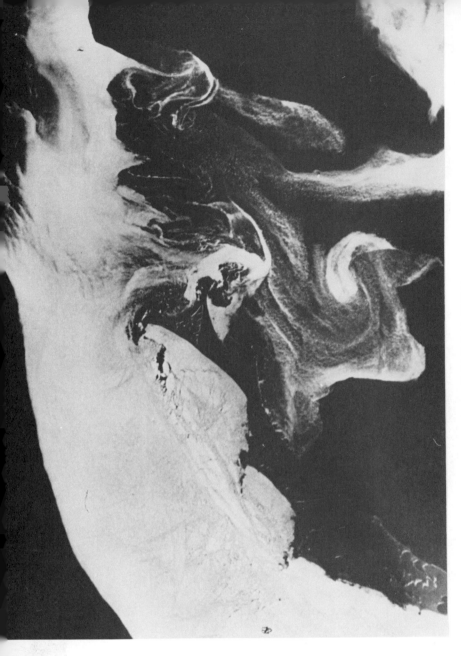

A near vertical view of the island of Sakhalin, USSR. It is located just off the Pacific coast of Russia and is north of the Japanese island of Hokkaido. Note the presence of much ice.

such huge numbers that they color the water greenish, brownish, or yellowish. Spotting color changes in the eddy off New Zealand, the astronauts knew that they had found a new fishing ground.

Off the east coast of South America, Carr's crew spotted eddies in the Falkland Current flowing northward along the coast of Argentina. Jerry described the current as "one of the most spectacular currents in the world. You see it very plainly as a light but brilliant—almost fluorescent—green." This color comes from life-giving plankton, and this area is a popular fishing ground for Argentinian and Soviet fishermen.

The greatest effect of the mapping of eddies by Skylab will be on our knowledge of weather. Such large pools of cold water could explain the way hurricanes behave. The energy of these storms comes from heat that rises into the air from the ocean. Hurricanes tend to become weaker when they pass over cold water. Thus, from the temperature of the ocean, forecasters can tell whether a hurricane or cyclone will increase or decrease in fury.

Large numbers of eddies could cool a lot of air and cause changes in weather. Weather and climate depend on the exchange of heat energy between the oceans and air. Eddies are more or less permanent features, so they must play an important role in this exchange. "These eddies," said Carr, "will answer a lot of questions people have been asking about the causes of change in weather and climate."

The visual observation program proved the usefulness of man as a scientific observer in space, Wilmarth said. Such observations will be part of most, if not all, future manned space missions, he believes.

"The EREP instruments used on Skylab probably could be designed so that there would be no need for man to operate them," Bill Schneider told me. "But this would require a very complex and expensive communication system between the ground and a satellite. Man is still necessary to make the best possible use of sensors—to use his judgment when to take data and when not to. He can take advantage of unexpected opportunities and situations. As we saw in Skylab, man can solve problems, make repairs and change the types of sensors that are used. Looking toward the future we can see that some space missions must be unmanned because of costs. But Skylab clearly has shown that there will be times when man is required up there."

The End of a Space Era—The Beginning of Another

On the forty-eighth day in orbit, reporters asked the Skylab astronauts many of the questions that you might ask them if given the chance.

"What do you miss most besides your families?"

"Good food and the ability to eat anytime you want to," replied Pogue.

"I miss football and a cold can of beer while watching the game," said Carr. "I miss the opportunity to just sit down and relax—to listen to music, or stare out the window and gather my thoughts."

Gibson agreed with Carr. The thing he missed most was the chance to relax at the end of the day.

One reporter asked Pogue about the mistakes he had made, such as ruining data on nine EREP passes by forgetting to put filters on some cameras. After the incident when he tried to conceal his vomiting, Pogue said, he tried too hard to do a good job. Instead, "I proceeded to make more errors and to berate myself," he told the reporter. "Finally, I realized that I'm a human being, that I'm going to make mistakes, and I have to accept myself for what I am."

This led him to a new attitude about life. "When I see other people now," he said, "I try to see them as human beings and to put myself into the human situation instead of trying to operate like a machine. I tried to operate like a machine and I was a gross failure. Now I'm trying to operate like a human being within the limitations I possess."

"I feel the same way Bill does," Carr added. "People in our line of work—

very technical work—are inclined to move along with blinders on. This mission is going to do me a lot of good in that it's going to increase my awareness of what else is going on besides what I'm doing."

Ed Gibson spent a lot of time on the mission looking at the Sun and stars. This gave him the feeling that life may exist in other parts of the universe besides our own. "When you're up here," he explained, "you see the Sun as one star. You see all the other stars, and you realize the number of possible combinations that could create life. It makes it [life on other planets] seem much more likely."

Getting Skylab Ready for Visitors

Reporters asked about the troublesome gyros. Did the astronauts think they could stay up for the planned 84 days?

"The two gyros we have left seem to be reasonably stable," answered Carr. We've only had one or two instances where we used up more fuel than we figured we'd use. I'm thankful that the first two crews were efficient with their fuel usage, so that they left us a bagful. Right now, I think, by golly, we're going to make it."

Then on January 31, Carr boasted: "We've been able to do the most extensive maneuvering of any of the missions, and on only two gyros." Next day, flight director Don Puddy said: "There's no reason now to believe that the problem gyro is not going to hang in there with us."

Astronaut Edward Gibson has just emerged from the Skylab door on hatchway. Carr was above on the Apollo Telescope Mount when he shot this photo during the final Skylab extravehicular activity. Astronaut Pogue remained inside.

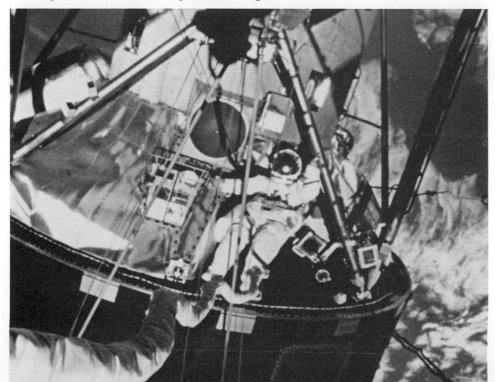

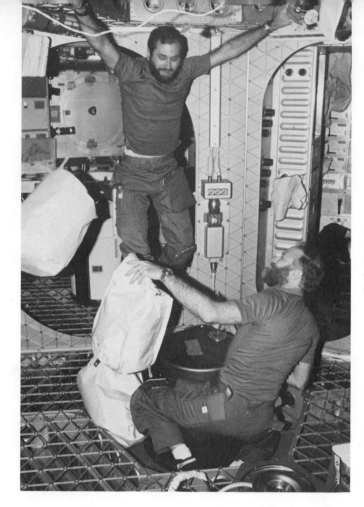

William Pogue (standing) and Gerald Carr (kneeling) passing trash bags through the trash airlock in the crew quarters.

The crew did the last EREP work on February 1, and took the last photos of the Sun next day. On Sunday, February 3, the final spacewalk took 5 hours and 19 minutes. Film and cameras were removed from the telescope mount for the last time. The astronauts completed the material processing experiments on Tuesday. They performed the last medical test on Wednesday, then they did a weightless dance to celebrate. Laughing and grinning, the three of them turned somersaults and did backflips. For a finale, each went to a different place in the spacious upper floor of the workshop. They leaped off into each others arms, hugging one another and spinning around.

That same day, as Skylab passed over Egypt, the crew fired four jets aboard the Apollo spacecraft. The jets burned for three minutes and pushed the space station into an orbit about 6 miles higher, or up to 276 miles. (The new orbit went from a high point of 283 miles to a low point of 269 miles.) This reduced the drag of the thin atmosphere so that the station would remain aloft as long as ten years after the astronauts left. The crew also fixed the "front door," or docking hatch, so that the station could be revisited in the future. Even if a visiting spaceship could not dock with the station, astronauts, or cosmonauts, could don space-suits, walk over from their ship and open the door.

Once inside, they would find a "time capsule" prepared by Carr, Gibson, and Pogue. This capsule contained samples of bread, pudding, beverages and flame-resistant cloth, a fan, eight camera filters, electrical cables, a flight-data file, star chart, fire-detection panel, a surgical glove, and a control unit. No plans existed for a revisit, but NASA decided to make one possible. Soviet cosmonauts might pay Skylab a visit in their three-man Soyuz spacecraft. Or one of the U.S. missions planned for the 1980s might include a stop at Skylab. Then the visitors would be able to examine the objects in the time capsule and see how they had been affected by drifting for years in space.

The Last Splashdown

On Thursday, February 7, the astronauts were awakened at 4 A.M. They worked until 1 P.M. packing film and data in the Apollo spacecraft, and closing up the station. They went to bed at one o'clock, and controllers woke them again at 8:40 that night. Flight planners designed this routine to give the astronauts maximum sleep and to bring their spaceship down as close as possible to the west coast of the United States.

By 1:30 on the morning of February 8, everything had been turned off and 1817 pounds of "luggage" packed in the Apollo ship. Ed Gibson couldn't resist getting a final three minutes of data from the Sun on video tape before he shut down the telescope mount control panel.

The crew received a GO for undocking. At 5:34 A.M., over the Atlantic in the vicinity of Bermuda, Carr released the spring-loaded docking mechanism. But as it pushed them away, one of the capture latches caught the spacecraft. Carr had to fire some small thruster rockets to back away. Then he flew around the station for a last look.

"A lot of tender loving care has gone into this thing," Gibson commented. They saw that the sun shields rigged up by the first two crews had been burned brown by the Sun. "It's been a good home. I hate to think we're the last guys to use it," the scientist-astronaut said.

At 6:32 A.M., the big engine at the rear of the Apollo service module was fired, and the spacecraft dropped into a lower orbit. Another firing of the main engine at 9:36 sent the spacecraft on a long, controlled plunge downward from the skies above Southeast Asia toward the California coast.

Meanwhile, a leak broke out in the fuel system aboard the Apollo command module. Ground controllers quickly told the crew to open some vents in the fuel compartment to prevent poisonous gases from leaking into the cabin. They also warned the astronauts to put on oxygen masks if they smelled an acid or fishy odor. The smell would mean that poisonous fumes were getting into the cabin.

On top of the leak came what Carr later called "the great reentry crisis." It was time to separate the command and service modules. Part of the procedure called for pulling four circuit breakers. "In the fury of getting all our marbles gathered up," Carr said, "we pulled the wrong four breakers. When we sepa-

The command module containing Skylab astronauts (Carr, Gibson and Pogue) splashed down in calm Pacific seas at 11:17 a.m., EDT, February 8, 1974, 283 kilometers (176 statute miles) southwest of San Diego, California, at completion of the 84-day space mission.

U. S. Navy para-rescuemen jump from helicopter during Skylab recovery operations in the Pacific Ocean.

rated, we had that long moment of silence as I twitted the hand controller and nothing happened. Our hearts fell and our eyeballs popped. But we immediately moved to a backup system to fire the command module thusters. There was an instant of stark realization that something was amiss. But thanks to good design we had plenty of backup, and it was no problem at all."

At 10:01 the spacecraft plunged back into the Earth's atmosphere. The astronauts were now about 1,700 miles west of Portland, Oregon, and traveling better than 17,000 miles an hour. The fiery envelope of electrified air around their spacecraft blacked out all radio communication for about three minutes. When the blackout ended, the spacecraft had reached a point about 300 miles west of Los Angeles.

Ten minutes after reentry the small drouge parachutes opened. Then the large orange and white main parachutes ballooned out and slowed their descent to Earth.

"What a beautiful sight!" Gibson exclaimed as he looked up at the fully opened chutes.

They hit a calm sea at 10:17 A.M., some 175 miles southwest of San Diego and about 3 miles from the recovery carrier U.S.S. *New Orleans*. It was Friday, February 8.

"Welcome home," radioed the *New Orleans*.

"Glad to be back," replied Carr, Gibson, and Pogue.

Man's longest space journey was over.

Shaky But Smiling

In about forty-five minutes, the spacecraft, with the astronauts inside, was hoisted aboard the carrier. The crew had smelled no fumes, and no sign existed of the leak in the fuel system.

The astronauts, smiling and happy, were obviously in good spirits. Carr and Pogue had grown big, bushy beards, but Gibson kept clean-shaven. As they came out of the spacecraft, they waved to everyone and said they "felt great." But they were so wobbly they had to be helped into chairs mounted on a platform just outside the Apollo ship. A forklift truck raised and lowered the platform and took Carr, Gibson, and Pogue to the special medical laboratory where doctors would examine them.

Back in Houston, at the Mission Control Center, everyone shook hands and puffed cigars. On a large screen in the center of the control room, a display showed the insignia of each of the crews. Superimposed on this, controllers put the words: "Skylab, the world's largest space station, man's longest venture into -space, improved understanding of the universe, rediscovery of the planet Earth, accomplishment of major vehicle repair—man, machine and spirit in a truly great accomplishment."

At a meeting of NASA officials and newsmen, Skylab director Bill Schneider went over the achievements of the last and longest mission. In 84 days, 1 hour, and 15 minutes, Carr, Gibson, and Pogue circled the Earth 1,213 times and traveled 34,508,814 miles. Carr and Pogue took Skylab's longest spacewalk—7 hours, 1 minute. During four spacewalks on the mission, the crew spent 22 hours, 21 minutes outside the space station.

"The real payoff—the reason for the whole Skylab project—is the data we've obtained," said Schneider. "We planned thirty Earth resources passes on this mission, and we did thirty-nine. This yielded 20,500 pictures of our planet and nineteen miles of data on tape. On the corollary experiments, such as material processing, student investigations and maneuvering-unit testing, we planned twenty-eight experiments and we accomplished twenty-eight."

In the field of Sun watching, Ed Gibson became the first man in history to

Spacecraft was hoisted abroad the U.S.S. New Orleans following the third and final manned mission in the Skylab program, designed to gain knowledge in space for improving life on Earth.

see and record a solar flare from its beginning to its end. After much patient waiting, he made the historic observation on January 21, 1974. "Thanks to Dr. Gibson's alertness," commented solar scientist Allen S. Krieger, "we have been able to observe, for the first time, a process whereby energy is transferred from the Sun's magnetic field into heat energy that reaches our planet. If we can further unlock the secrets of this energy-transfer process, it may be possible to develop processes on Earth that will result in cheaper sources of energy."

Shrinking, Shifty Spacemen

Meanwhile, on the carrier U.S.S. *New Orleans,* the astronauts went through six and a half hours of medical tests on Friday afternoon. Dr. Royce Hawkins reported that they were unsteady on their feet and experiencing some vertigo —the feeling that you, or everything around you, is whirling or moving. They also felt tired. But their condition improved later in the day, and the unsteadiness and vertigo were gone by the time they flew home to see their families on Sunday. Dr. Jerry Hordinsky, head of the medical team on the carrier, reported that the three tested out to be in "as good or better shape" than the first two crews.

Where it had taken 21 days for the hearts and muscles of the first crew to get back to normal, Carr's crew was back in much the same condition in about five days. And this despite the fact that they spent three times longer in space, 84 as compared to 28 days. The calves of all the crews shrank as much as one and a half inches in diameter during their flights. The legs of Bean's crew, who were up for almost 60 days, took 21 days to return to normal. For Carr's crew, it took 11 days.

Surprisingly, then, the longer a crew stayed in space the better physical shape they were in when they returned. How come? "Exercise played a major role," said Hawkins. "This and having enough time for the body to adjust to weightlessness."

Pete Conrad's crew exercised a half hour each day. When Conrad recommended this be increased, Bean's crew went to an hour each day. When they came back, the second crew recommended that the daily exercise period be increased again. Carr's crew worked out one and a half hours each day. In addition to the wheelless bike that the other crews had, they used a treadmill. This device was designed by scientist-astronaut William E. Thornton to exercise leg muscles not used during other activities.

"If we hadn't exercised, we'd have been like jellyfish when we came back," Pogue told the doctors.

Hawkins pointed out that the astronauts showed little change in their physical condition after about forty days in space. Therefore, the bodies of the first crew did not undergo all the changes needed to become adapted to weightlessness. The second and third crews did stay up long enough to reach a

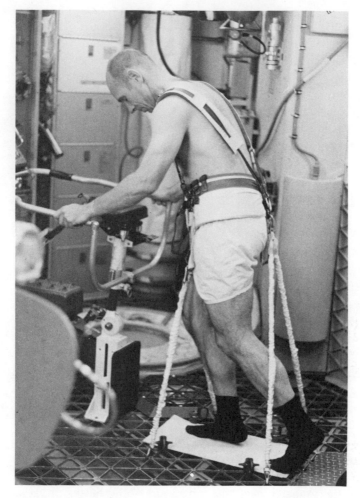

Astronaut tests a tread-mill-like exercise device developed for maintaining the leg and back muscles of crewmen on third Skylab flight. First two crews had no exercise device on board capable of adequately maintaining these muscles.

stable level. As long as they exercised, too, they had less trouble than Conrad's crew readapting to Earth's gravity. It took the hearts and muscles of the second and third crews about the same time to return to normal—five days. Both crews probably could have stayed up much longer and not taken much more time to return to their preflight condition.

A big surprise was the finding that astronauts did not change as much as doctors thought they would. "We didn't see any appreciable amount of muscle loss," Hawkins told me. "Losses in bone calcium were not enough to cause bone weakness or breaks. The astronauts' hearts didn't deteriorate, or get 'lazy.' Before we flew Skylab, I thought we would see greater changes in the heart and blood vessels."

The biggest changes occurred during the first few days after splashdown, as the astronauts tried to readapt to Earth's gravity. Their hearts had to work harder to pump the same amount of blood. This happened for two reasons. First, their blood tended to shift downward, or pool in their legs, when they first came back under the influence of gravity. Second, they lost some of their blood, both red blood cells and plasma, while in space. This loss both made them tired and forced their hearts to work harder.

The loss of blood cells and plasma is part of adapting to zero gravity, Hawkins and Hordinsky believe. You don't need as much fluid or red cells in zero gravity, so the body does not produce the same amounts as it would on Earth.

Shifts in body fluids caused by the off-and-on gravity produced other weird effects. With less fluid to pump, the hearts of the astronauts became smaller, shrinking in diameter. Before Skylab, doctors feared that a wasting away of the heart muscles produced this shrinkage. But careful examination of the third crew revealed that the chambers of their hearts got smaller simply because less blood circulated through them.

Fluid loss also accounted for much of the weight drop experienced by the astronauts—an average of about five pounds.

Upon first getting into space, astronauts reported a feeling of fullness in their heads. Television photos showed a rounding of their faces and bulging of veins in the neck. This is produced by an upward movement of blood as the pull of gravity decreases. Pooling in upper body also caused their chests to expand and the lower parts of their legs to shrink. This is why the astronauts' calves got smaller.

The most startling effect of fluid shift was to make astronauts taller. Measurements made aboard Skylab showed that Carr, Gibson, and Pogue gained as much as 2 inches in height. "The same thing must have happened on the other missions, but we didn't find it," admitted Hordinsky.

The downward pull of gravity keeps your backbone compressed. When this force was removed, the cushions, or disks, between the bones of the astronauts' spines expanded. The backbone increased in length, and they "grew" taller. When they returned to Earth, gravity pulled the bones of their backs closer again, and the astronauts shrank to normal size. Bill Pogue is normally 5 feet, 9 inches tall. But when doctors measured him on the carrier, he was almost 5 feet, 10 inches. He and the other astronauts returned to their normal height in less than two days.

Hawkins called the medical results of Skylab "remarkable" and "most encouraging for the future of manned spaceflight." He said that Skylab not only opened the way for astronauts to make longer spaceflights but for non-astronauts to fly in space.

"We have found that the human body can adapt to zero gravity," explained Hawkins. "This means that the average healthy person can probably go on spaceflights of up to a month. Anyone can probably fly in space as long as he does not have heart, blood pressure, or other medical problems, and as long as he exercises at least one hour daily. This would include women as well as men."

"Telemetry Off"

The Skylab program did not end with the departure of the crew. After the last astronauts left, ground teams performed final tests on some of the equipment. They did things they didn't dare do with men aboard, such as dumping all the instructions in a computer brain, then seeing if they could put them back in again. Engineers tried to start up the one gyro that failed, but they had no luck.

Finally, on Saturday, February 9, they placed Skylab in a vertical position with the docking hatch pointing away from Earth. In this position it would be subjected to the least amount of drag and so would be less likely to start tumbling. The telescope mount faced out into space, viewing now with blank eyes the universe whose secrets it probed for nine months.

At 2 P.M., the last command was sent to man's first home in space—"telemetry off."

A Look Toward the Future

NASA had planned to operate the Skylab for 240 days. It actually operated for 271 days. The space agency wanted Skylab to be manned for 140 days. The three crews actually spent almost 172 days aboard, and they orbited the Earth 2,476 times. The exact total of time man spent on the first space station came to 171 days, 13 hours, 15 minutes and 25 seconds. That's more time than all the previous U.S. manned spaceflights put together. The combined total of the Mercury, Gemini, and Apollo Moon-landing programs came to 146 days, 21 hours, 36 minutes and 8 seconds.

"We thought we were being quite bold when we planned the Skylab missions," said Schneider. "We scheduled more work than we expected the astronauts to complete. We planned 565 hours of taking data on the Sun, but we actually got 755 hours. We planned 701 hours of medical experiments, we obtained 822. We thought we'd be lucky if we got 60 Earth observation passes, we did 90. We had 10 materials processing experiments planned, we did 32. We did more student experiments than we thought we'd be able to do.

"Each scientist has his own idea of what the most important results were. Certainly we changed the whole theory of what goes on in the Sun and how our world is heated. The materials processing experiments indicate that we can make commercially valuable products in space. We thought man was going to be able to do some work up there. But we were completely surprised to find that, given the proper restraints [against drifting around in weightlessness], the proper tools, and the proper training, there is very little a man can't do in space that he can do on Earth."

How much did Skylab cost and what benefits did it produce? NASA director Dr. James C. Fletcher said that the total cost came to $2.6 billion, or about one-tenth of the Moon-landing program. According to Fletcher, what we bought for that was a continuation of U.S. leadership in manned spaceflight and proof

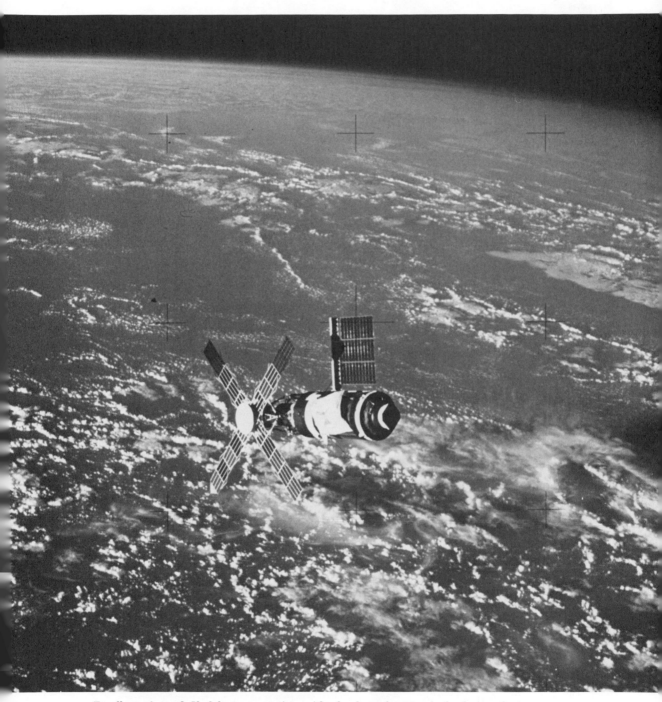

Excellent view of Skylab space station with clouds and water in background. Basically an experimental space station, Skylab confirmed that we are on the right track in proceeding to develop the Space Shuttle.

of man's ability to live and work effectively in space. We clearly demonstrated that man could perform valuable services in Earth orbit as observers, scientists, engineers, and repairmen. We obtained a wealth of new scientific information that could not be obtained from observatories on the ground.

Equally important, Fletcher said, Skylab paved the way for spaceflights of the future. The three missions strengthened the Soviet Union's interest in cooperating with NASA, and smoothed the way for the U.S.-Russian joint spaceflight in 1975.

"Skylab confirmed that we are on the right track in proceeding to develop the Space Shuttle," Fletcher continued. The shuttle is a reusable space vehicle that will be launched as a rocket, but will return to Earth and land like an airplane. Expected to begin space operations in the 1980s, this vehicle would carry both U.S. and European astronauts on regularly scheduled spaceflights. With a regular schedule and vehicles that can be used again, space travel is expected to be much cheaper than in the 1960s and '70s.

The shuttle will carry as passengers American and European scientists and technicians, including women. They will operate instruments and equipment to do space manufacturing, and study the Earth, Sun, and universe. Later shuttle missions will launch and repair unmanned satellites and may even launch men on flights to distant planets. Many of the things that were learned about living in space on Skylab are going into the planning of these missions. They range through everything from the number of windows needed and basic eating and housekeeping methods to knowledge of the best scientific instruments to do a particular job.

"Skylab was basically an experimental space station," remarked Fletcher. "However, it possessed many of the ingredients that will characterize operational missions of the future. It has moved the space program from the realm of the spectacular to an almost businesslike routine. It has contributed to an orderly transition from the Moon landings of the 1960s to the use of space for the benefit of man in the 1980s and 1990s."

Thus, Skylab marked a turning point from short to long space missions. From small to large spacecraft. From the age of space discovery and exploration to the age of space colonization. From daredevil test pilots proving that man can survive in space to scientists, engineers, technicians, and even tourists, living and working beyond Earth on a routine basis.

Index

Page references to illustrations are in *italic type*.